Polymers from Plant Oils

Scrivener Publishing
100 Cummings Center, Suite 541J
Beverly, MA 01915-6106

Publishers at Scrivener
Martin Scrivener (martin@scrivenerpublishing.com)
Phillip Carmical (pcarmical@scrivenerpublishing.com)

Polymers from Plant Oils

2nd Edition

Alessandro Gandini
and
Talita M. Lacerda

Scrivener
Publishing

WILEY

This edition first published 2019 by John Wiley & Sons, Inc., 111 River Street, Hoboken, NJ 07030, USA and Scrivener Publishing LLC, 100 Cummings Center, Suite 541J, Beverly, MA 01915, USA

For more information about Scrivener publications please visit www.scrivenerpublishing.com.

Wiley Global Headquarters
111 River Street, Hoboken, NJ 07030, USA

For details of our global editorial offices, customer services, and more information about Wiley products visit us at www.wiley.com.

Library of Congress Cataloging-in-Publication Data
Names: Gandini, Alessandro, author. | Lacerda, Talita Martins, author.
Title: Polymers from plant oils / Alessandro Gandini and Talita M. Lacerda.
Description: 2nd edition. | Hoboken, NJ : John Wiley & Sons, Inc., 2018. |
Includes bibliographical references and index. |
Identifiers: LCCN 2018044842 (print) | LCCN 2018046235 (ebook) | ISBN 9781119555827 (ePub) | ISBN 9781119555339 (ePDF) | ISBN 9781119555797 (hardcover)
Subjects: LCSH: Vegetable oils--Industrial applications. | Plant polymers. | Polymerization. | Biomass chemicals. | Renewable natural resources.
Classification: LCC TP680 (ebook) | LCC TP680 .G28 2018 (print) | DDC 665.3--dc23
LC record available at https://lccn.loc.gov/2018044842

Cover images: Provided by the authors
Cover design by: Russell Richardson

Set in size of 11pt and Minion Pro by Exeter Premedia Services Private Ltd., Chennai, India

10 9 8 7 6 5 4 3 2 1

Contents

Preface to First Edition

Vegetable (or plant) oils, shown here as plant triglycerides, constitute one of the most abundant variety of renewable resources on our planet. They have been exploited by humans for millennia for three major purposes: *food and feed*, *energy sources* and precursors to useful *products and materials*. Ancient utilisations and applications (empirical by definition) contributed to ensure: daily sustenance for people and animals; comfort through heating and illumination; development of protective, writing and artistic coatings in the form of film-forming materials that 'dried in air'. They were then adopted as pristine natural compounds, though in some instances other components were mixed with them, as in the case of inks and lacquers. Simple chemical transformations carried out empirically were also developed, as in the manufacture of soaps simulating the process applied to animal fats.

Unraveling of their chemical structure, and hence understanding of their reactivity, favoured more rational processing and widened the range of applications throughout the twentieth century. Another more important 'revolution' has begun at the start of the third millennium with expansion of the research/development of biofuels and macromolecular materials.

This book is devoted *exclusively* to the latter realm, with particular emphasis on recent trends, progress, achievements and perspectives, with broad treatment of the subject, including inks, paints and coatings, in addition to the more conventional bulk thermoplastic and thermosetting polymers.

In the field of film-forming materials, use of alkyd resins incorporating plant oils or their derivatives has been a standard practice for a century, but no major qualitative advance was introduced until recently. The same applies for bulk polymers based on vegetable oils, of which linoleum (first commercialised in the middle of the nineteenth century) was for a long time the only important representative of these materials. Nylon 11 (commercialised under the name of Rilsan based on castor oil as a precursor) has been an important addition to this small family from the 1950s onwards. In other words, vegetable oils represented a very modest presence as basic constituents of macromolecular materials up to about a decade ago, but the situation has evolved radically since then.

The purpose of this book is to highlight this impressive and promising ongoing trend, which is also occurring in all other areas of the novel burgeoning domain of polymers from renewable resources.

We wish to thank most heartedly Joan Gandini for her constant help in improving the language and style of the manuscript. The authors kindly acknowledge FAPESP for T.M.L.'s post-doctoral fellowship (2012/00124-9) and CNPq for A.G.'s visiting professorship (Science Without Borders programme, PVE 401656/2013-6).

February 2015

Preface to the Second Edition

The publishers of the first edition of this book (2015), Smithers Rapra, relinquished their copyright to the authors who prepared the present second edition for publication under the auspices of Scrivener Publishing. There are no *qualitative* changes in the contents of this edition with respect to the first, which we still considered quite appropriate, whereas our attention focused on bringing up to date a number of issues related to recent relevant facts and important contributions. We feel that the topic of the book has maintained a high level of interest, as judged by the continuing flow of publications and new materials being developed by both academic and industrial research activities throughout the world. We trust therefore that the message contained in the original preface remains fully justified.

The authors kindly acknowledge FAPESP for financial support (2017/16062-6).

Alessandro Gandini and Talita M. Lacerda
August 2018

1 Introduction

1.1 Setting the Stage

World production of major oilseeds has increased from 331 million tonnes to 573 million tonnes in the last decade, whereas the harvested area has increased from 186 million acres to 278 million acres in the same period. According to the United States Department of Agriculture (USDA), soybean oilseeds represent over half of the total production of oilseeds and are mainly grown in Brazil, USA and Argentina; followed by rapeseed (grown in European Union (EU)-27, Canada and China), cottonseed (China and India) and sunflower oilseeds (Ukraine, EU-27 and Russia). **Figure 1.1** shows the total production of major oilseeds around the world in 2017/2018.

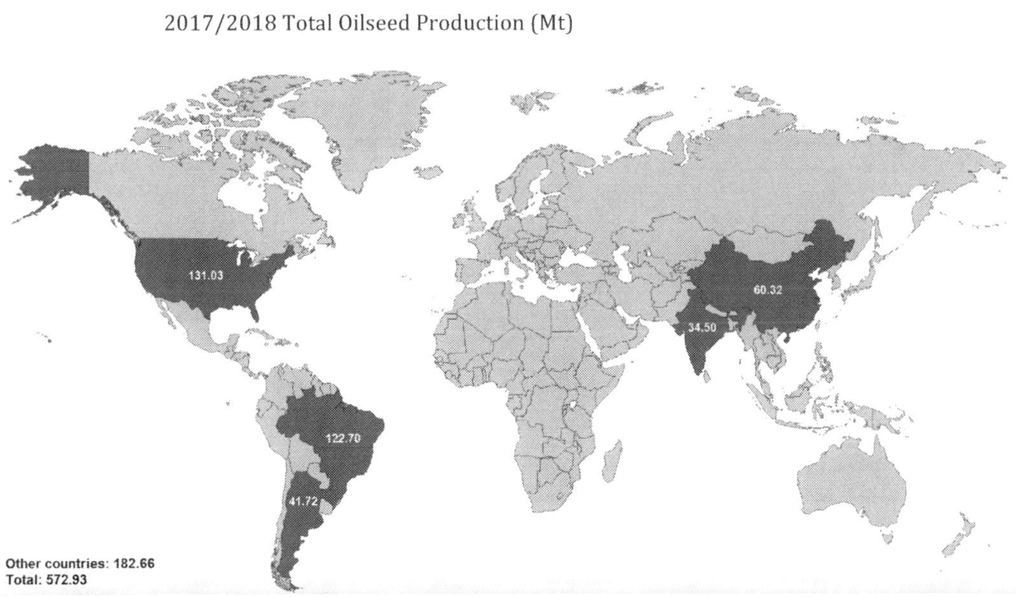

Figure 1.1 Total production of oilseeds 2017/2018 (Mt). Adapted from United States Department of Agriculture *http://www.fas.usda.gov/psdonline* ©United States Department of Agriculture [1]

The figures shown above translate from 96 million tons (Mt) in 2002/2003 to 199 Mt in 2017/2018 of actual oils, with an average growth rate of ~7 Mt a year, (**Table 1.1**) [1]. This increasing demand is associated with the needs to feed an increasing population and, more recently, to the demand for biodiesels as partial replacements for fossil fuels. Minor vegetable oils such as castor oil and linseed oil are almost solely used for industrial applications because they are not appropriate for consumption by humans or animals.

Table 1.1 Major vegetable oils: world supply and distribution (Commodity view) – (million metric tonnes) [1]

Oils	2002/2003	2017/2018			
	Production	Production	Imports	Exports	Consumption
Coconut	3.16	3.54	1.50	1.69	3.25
Cottonseed	3.51	5.09	0.05	0.09	5.02
Olive	2.51	3.27	0.88	1.00	2.86
Palm	27.71	69.72	47.23	50.15	65.17
Palm kernel	3.36	8.15	2.75	3.19	7.55
Peanut	4.62	6.00	0.24	0.28	5.98
Rapeseed	12.21	28.75	4.34	4.61	29.07
Soybean	30.57	55.81	10.23	10.71	55.43
Sunflower seed	8.12	18.35	8.10	9.54	17.09
Total	**95.77**	**198.68**	**75.31**	**81.25**	**191.42**

Palm oil and soybean oil are the most important (as well as the most widely exported) oils, followed by rapeseed and sunflower counterparts (**Table 1.1**).

The common structure of vegetable oils discussed here is that of aliphatic triglycerides (**Scheme 1.1**), in which the 'fatty acid chains' R_1, R_2 and R_3 are most often identical, but can also vary, within a given molecule. The length of the fatty-acid chain is 14–22 carbon atoms, but most members bear 16 or 18 units. The other important feature of these linear aliphatic motifs is the possible presence of C=C unsaturations, which range from 0 to 3. More than 1,000 fatty acids have been identified, but only »20 are present in appreciable quantities in vegetable oils [2, 3].

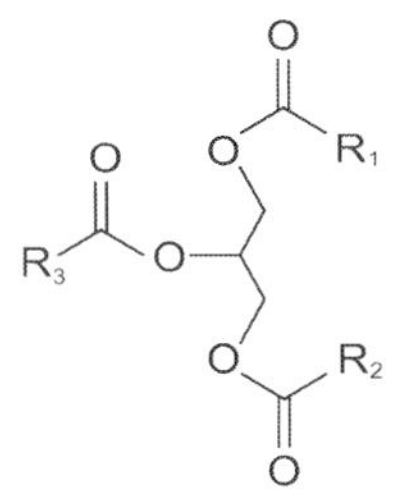

Scheme 1.1 Generic structure of a natural triglyceride component of vegetable oils in which R_1, R_2 and R_3 are fatty-acid chains

Vegetable oils comprise a mixture of triglycerides (albeit with one or two specific structures which usually predominate). These compositions vary according to plant species, crop type, season, and growing conditions [4]. **Table 1.2** enumerates the most common fatty acids in the triglycerides of plant oils and **Scheme 1.2** shows their structures.

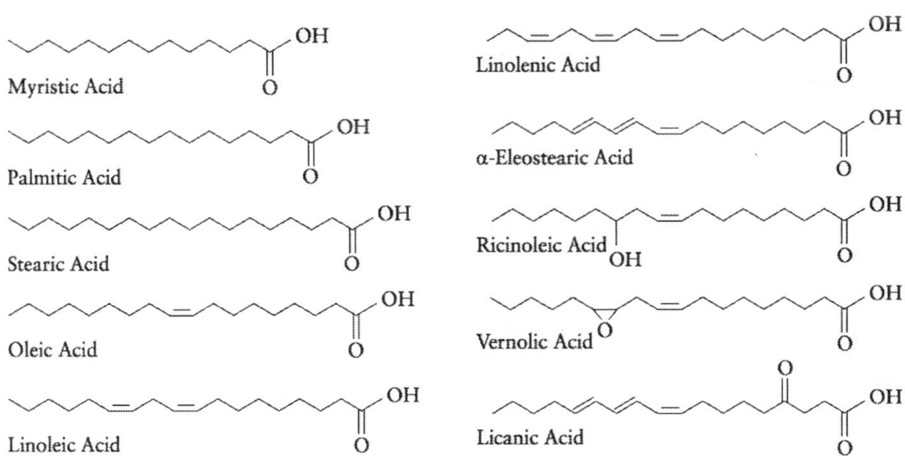

Scheme 1.2 Structures of the most common fatty acids

Table 1.2 Most common fatty acids in vegetable triglycerides [5, 6]

Trivial name	Systematic name	Structure (C:DB)*	Formula
Lauric acid	Dodecanoic acid	12:0	$C_{12}H_{24}O_2$
Myristic acid	Tetradecanoic acid	14:0	$C_{14}H_{28}O_2$
Palmitic acid	Hexadecanoic acid	16:0	$C_{16}H_{32}O_2$
Stearic acid	Octadecanoic acid	18:0	$C_{18}H_{36}O_2$
Arachidic acid	Eicosanoic acid	20:0	$C_{20}H_{40}O_2$
Behenic acid	Docosanoic acid	22:0	$C_{22}H_{44}O_2$
Lignoceric acid	Tetracosanoic acid	24:0	$C_{24}H_{48}O_2$
Palmitoleic acid	*Cis*-9-hexadecenoic acid	16:1	$C_{16}H_{30}O_2$
Oleic acid	*Cis*-9-octadecenoic acid	18:1	$C_{18}H_{34}O_2$
Linoleic acid	*Cis,cis*-9,12-octadecadienoic acid	18:2	$C_{18}H_{32}O_2$
Linolenic acid	*Cis,cis,cis*-9,12,15-octadecatrienoic acid	18:3	$C_{18}H_{30}O_2$
a-Eleostearic acid	*Cis,trans,trans* -9, 11,13-octadecatrienoic acid	18:3	$C_{18}H_{30}O_2$
Erucic acid	*Cis*-13-docosenoic acid	22:1	$C_{22}H_{42}O_2$
Ricinoleic acid	12-Hydroxy-*cis*-9-octadecenoic acid	18:1	$C_{18}H_{34}O_3$
Vernolic acid	12,13-Epoxy-*cis*-9-octadecenoic acid	18:1	$C_{18}H_{32}O_3$
Licanic acid	4-Oxo-*cis,trans,trans*-,11, 13-octadecatrienoic acid	18:3	$C_{18}H_{28}O_3$

*C indicates the number of carbon atoms and DB the number of double bonds in the fatty-acid chain

Some fatty acids (e.g., lauric, myristic, palmitic, stearic) are saturated, whereas others are monounsaturated (e.g., oleic, erucic) or polyunsaturated (e.g., linoleic, linolenic). In most vegetable oils, the double bonds of the fatty-acid chains are in the *cis* configuration (e.g., oleic, linoleic), although *trans* counterparts may also be present (e.g., a-eleostearic, licanic). The double bonds are more often non-conjugated (e.g., in linoleic and linolenic motifs) but conjugated sequences are also encountered (e.g., in eleostearic and licanic structures). Some oils contain fatty acid esters with other moieties along their chains, such as ricinoleic, vernolic and licanic structures with hydroxyl, epoxy and carbonyl groups, respectively.

Isolation of vegetable oils from their seeds is carried out mechanically or by solvent extraction [7]. The mechanical process consists of submitting the beans, cells and oil bodies to shearing to liberate oil. Heat is generated during this procedure, which can induce a negative effect on the proteins therein. Advantages of the mechanical-isolation process reside in its low cost, low investment and safety in terms of environmental concerns because it does not involve solvents or hazardous substances. It is, nevertheless, marred by poor yields of oil extraction because the amount of oil left in the ensuing residues can be £7%.

The principle of this type of solvent extraction is based on diffusion of solvent through seeds and subsequent solubilisation of oil. The most common solvents used in this process are alkanes with low boiling points such as hexane. The key parameter of this process is the rate of diffusion of the solvent into the oil body. This process is more efficient than its mechanical counterpart but involves use of volatile organic solvents (though their recuperation is highly optimised).

After isolation, vegetable oils are refined to obtain high-quality products free from impurities such as phosphatides, free fatty acids, gummy substances, coloured bodies, tocopherols, sterols, hydrocarbons, ketones, and aldehydes [2].

Composition of vegetable oils is highly variable as a function of the associated species, which determines their possible applications as renewable feedstock. **Table 1.3** provides the typical composition of some vegetable oils in terms of their fatty-acid residues. In the case of the more exotic castor, oiticica and tung oils, the main fatty-acid residues are ricinoleic (87.5%), licanic (74%) and a-eleostearic acids (84%), respectively. These contents can, however, be modified by breeding or genetic modification of crops [8, 9]. For instance, erucic acid (43%) is the main fatty acid in standard rapeseed oil, but several of the rapeseed varieties in cultivation are based on zero content of erucic acid [10], given its toxicity in humans if ingested at high doses.

Physical and chemical properties of vegetable oils are dependent upon their fatty-acid composition. The numbers of double bonds (as well as their positions within the aliphatic chain) strongly affect oil properties. The actual number of carbon atoms comprising the aliphatic chains has a very minor role because most of these triglycerides have 18 of them and a few have 16. **Table 1.4** summarises some relevant properties of common vegetable oils and fatty acids. The average degree of unsaturation is measured by the iodine value (i.e., the amount of iodine (mg) that reacts with the double bonds of 100 g of a given oil).

Table 1.3 Average content of fatty-acid motifs in common plant oils

Oil	Fatty-acid motif					Average number of double bonds per triglyceride
	Palmitic	Stearic	Oleic	Linoleic	Linolenic	
Canola	4.1	1.8	60.9	21	8.8	3.9
Corn	10.9	2	25.4	59.6	1.2	4.5
Cottonseed	21.6	2.6	18.6	54.4	0.7	3.9
Linseed	5.5	3.5	19.1	15.3	56.6	6.6
Olive	13.7	2.5	71.1	10	0.6	2.8
Soybean	11	4	23.4	53.3	7.8	4.6
Tung	–	4	8	4	–	7.5
Castor	1.5	0.5	5	4	0.5	3
Palm	39	5	45	9	–	–
Oiticica	6	4	8	8	–	–
Rapeseed	4	2	56	26	10	–
Refined tall	4	3	46	35	12	–
Sunflower	6	4	42	47	1	–

Table 1.4 Some physical properties of triglyceride oils and fatty acids

Name	Viscosity (mPa.s)	Specific gravity	Refractive index	Melting point (°C)
Castor oil	293.4 at 37.8 °C	0.951 at 20 °C	1.473–1.480 at 20 °C	−20 to −10
Linseed oil	29.6 at 37.8 °C	0.925 at 20 °C	1.480–1.483 at 20 °C	−20
Palm oil	30.92 at 37.8 °C	0.890 at 20 °C	1.453–1.456 at 20 °C	33–40
Soybean oil	28.49 at 37.8 °C	0.917 at 20 °C	1.473–1.477 at 20° C	−23 to −20
Sunflower oil	33.31 at 37.8 °C	0.916 at 20 °C	1.473–1.477 at 20 °C	−18 to −16
Myristic acid	2.78 at 110 °C	0.844 at 80 °C	1.4273 at 70 °C	54.4
Palmitic acid	3.47 at 110 °C	0.841 at 80 °C	1.4209 at 70 °C	62.9
Stearic acid	4.24 at 110 °C	0.839 at 80 °C	1.4337 at 70 °C	69.6
Oleic acid	3.41 at 110 °C	0.850 at 80 °C	1.4449 at 60 °C	16.3

By convention plant oils are divided into three categories depending on their iodine values. Thus, oils are classified as *drying* if their iodine value is >130, *semi-drying* if it is 90–130, and *non-drying* if it is <90. Iodine values of some common fatty acids and their triglycerides are given in **Table 1.5**. This terminology is related directly to the aptitude of a film of oil to undergo oxido-polymerisation if exposed to the atmosphere, thereby turning from a viscous liquid to (eventually) a solid material. This reaction is induced by C=C moieties and hence the correlation between the degree of unsaturation of an oil (measured by its iodine value) and the susceptibility of its thin films to dry in air.

Table 1.5 Iodine values of some unsaturated fatty acids and their triglycerides

Fatty acid (number of carbon atoms)	Iodine value	
	Acid	Triglyceride
Palmitoleic (C16)	99.8	95
Oleic (C18)	89.9	86
Linoleic (C18)	181	173.2
Linolenic (C18) and α-eleostearic	273.5	261.6
Ricinoleic (C18)	85.1	81.6
Licanic (C18)	261	258.6

References

1. *http://www.fas.usda.gov.psdonline/circulars/oilseeds.pdf*
2. R.P. Wool and X.S. Sun in *Bio-Based polymers and Composites*, Elsevier, New York, NY, USA, 2005.
3. C. Scrimgeour in *Bailey's Industrial Oil and Fat Products*, John Wiley & Sons, New York, NY, USA, 2005.
4. M.A.R. Meier, J.O. Metzger and U.S. Schubert, *Chemical Society Reviews*, 2007, **36**, 1788.

5. M.N. Belgacem and A. Gandini in *Monomers, Polymers and Composites from Renewable Resources*, Elsevier, Amsterdam, The Netherlands, 2008.

6. B.K. Barnwal and M.P. Sharma, *Renewable and Sustainable Energy Reviews*, 2005, **9**, 363.

7. A. Gandini and N.M. Belgacem in *Monomers, Polymers and Composites from Renewable Resources*, Elsevier, Amsterdam, The Netherlands, 2008, p.42.

8. J.M. Dyer and R.T. Mullen, *Physiologia Plantarum*, 2008, **132**, 11.

9. A.S. Carlsson, *Biochimie*, 2009, **91**, 665.

2 Basic Chemical Notions

2.1 Drying Mechanism

Drying of unsaturated vegetable oils represents one of the oldest applications of these natural products. Film coatings have been exploited empirically for millennia to waterproof various surfaces, print images/texts, and to obtain beautiful images in paintings and lacquered objects. The actual details of this process in terms of chemical mechanisms and kinetics were unravelled in the middle of the 20th century, together with a reasoned choice of catalysts to accelerate it. **Scheme 1.1** provides a schematic view of this oxido-polymerisation for linoleic acid, which ultimately leads to crosslinked materials, a mechanism common to all similar structures, including atmospheric degradation of natural (polyisoprene) and synthetic (polybutadiene) rubber. The degree of unsaturation of a given oil (as assessed by its iodine value) is the key factor in determining the rate at which it will 'dry'. However, the C=C double bonds do not intervene directly in the first stages of the process (as shown clearly in **Scheme 2.1**) because they instead activate the C–H bonds next to them, upon which the oxygen molecule inserts itself to form a hydroperoxide moiety. This well-established chemical feature has, regrettably, not been assimilated by all authors. As a consequence, publications and textbooks occasionally illustrate the initial step of the process erroneously as occurring by the attack of oxygen onto the actual unsaturation. As with many other systems based on free-radical intermediates, this oxido-polymerisation attains a kinetic steady-state in which the rates of initiation and termination are equilibrated rapidly. **Scheme 2.2** is a simplified illustration of the overall kinetic pattern, again applied to linoleic acid.

This specific aspect of the polymerisation of plant oils has been discussed here because it constitutes a reminder of the oldest and best-known system that has been exploited for various applications related to coatings and thicker materials such as linoleum. That is, it represents well-established knowledge, as opposed to the different approaches employed in recent research aimed at developing novel macromolecular materials based on vegetable oils or their derivatives. These advances are the scope of the remaining sections of this book, which are organised according to the type of mechanism and nature of the ensuing polymer structures.

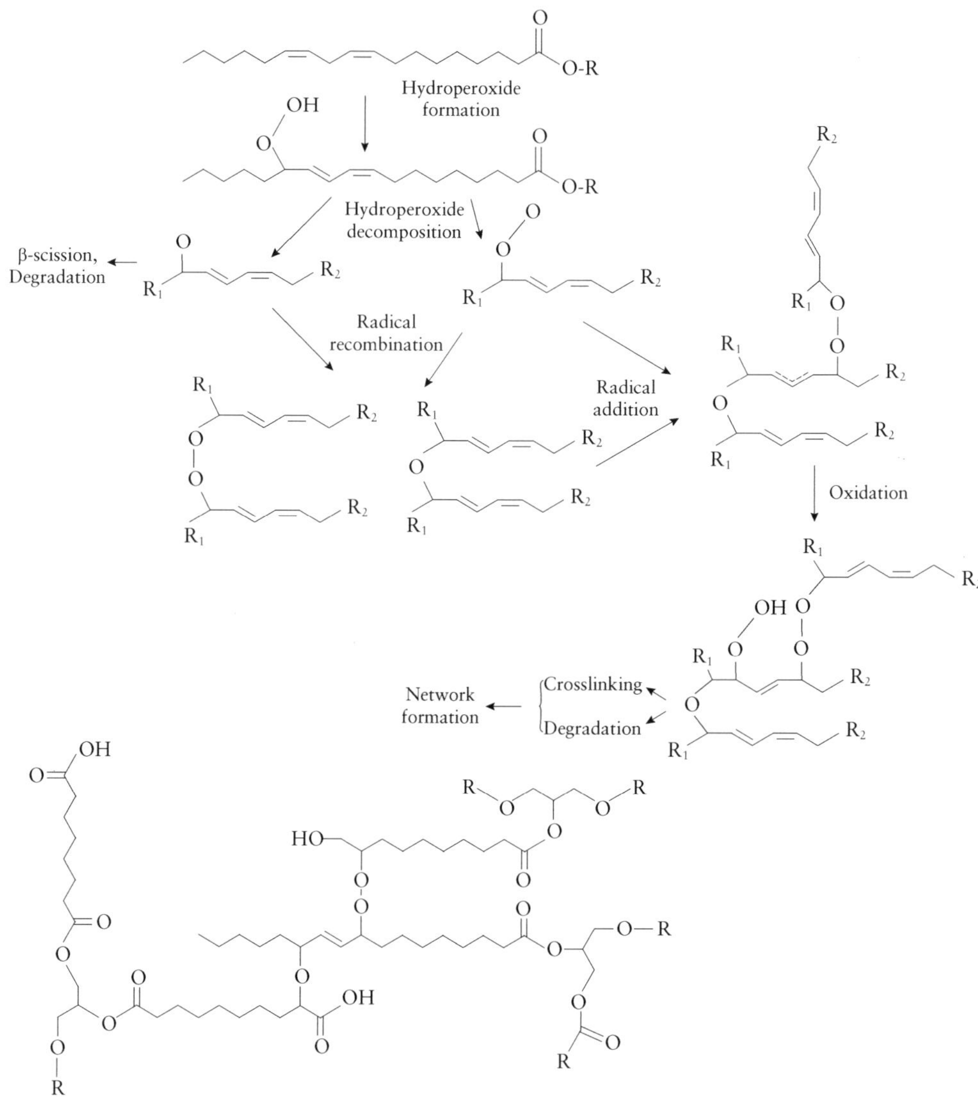

Scheme 2.1 Reactions involved in the oxido-polymerisation of unsaturated fatty acid and the corresponding oils

Before tackling the issues linked directly to macromolecular materials derived from plant oils, introduction to some basic notions related to the chemistry of triglycerides and their fatty-acid derivatives is crucial because they constitute the basic reactions that lead to monomers and polymers.

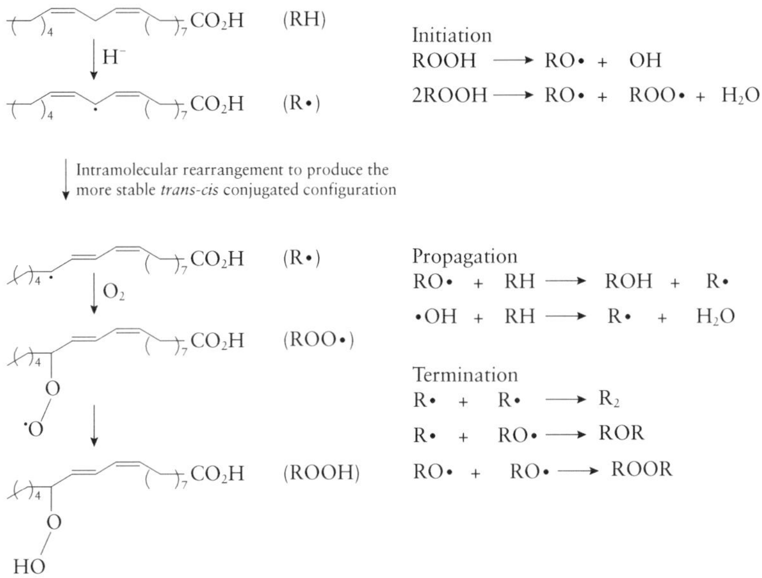

Scheme 2.2 Kinetic system associated with oxido-polymerisation

2.2 Reactive Sites

Considering the generic structure of a triglyceride incorporating C=C unsaturations, the sites susceptible to undergoing reactions or chemical modifications that may open fruitful ways of preparing polymers (directly or after preliminary conversions into suitable monomers) are shown in **Scheme 2.3** with respect to: 1) unsaturated bonds, 2) methylene moieties next to one of them (allylic carbons), 3) ester groups, 4) methylene sites next to one of them (alpha-carbons) and, to a lesser extent in terms of viability, terminal methyl groups (omega carbons). In addition to these general structural features, more specific oils incorporating such groups as hydroxyls or oxiranes, as described above, offer original alternative sources of chemical exploitation for polymer syntheses.

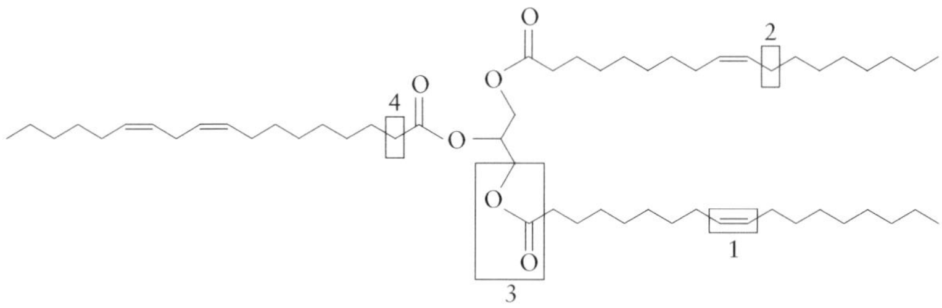

Scheme 2.3 Readily exploitable reactive sites in unsaturated triglycerides

The most important reactions related to ester and double-bond moieties have been the object of thorough reviews, particularly by Biermann and co-workers [1], Corma, Iborra and Velty [2], Behr and co-workers [3] and, within a wider scope, by Meier and co-workers [4]. The succinct treatment shown below is limited to transformations relevant to the scope of this book, *viz.* the routes to monomers or their precursors.

2.2.1 Reactions of the Ester Group

Transesterification (or alcoholysis) of plant oils, mostly using methanol, is the pathway for biodiesel production [5], which is the 'green' alternative to petro-diesel. This reaction generates three fatty-acid esters and glycerol. Biodiesel production increased from 900 mega litres (ML) to 15,000 ML from 2001 to 2009 and is expected to reach 45,000 ML by 2020 [6]. Alternative applications, such as 'green polymerisation solvents', have also been explored [7–9]. The glycerol ensuing from these processes constitutes an important precursor for the synthesis of many industrial commodities [2] to replace some mainstream fossil-based counterparts [10–12].

The basic principle of the transesterification reaction is based on the possibility of exchanging alkoxy groups if an ester is placed in a large excess of an alcohol in the presence of an acid or basic catalyst. The large excess of alcohol is important to shift the equilibrium to the side of the desired product. The mechanism of the transesterification reaction of a general ester and an alcohol in the presence of an acid catalyst is shown schematically in **Scheme 2.4**.

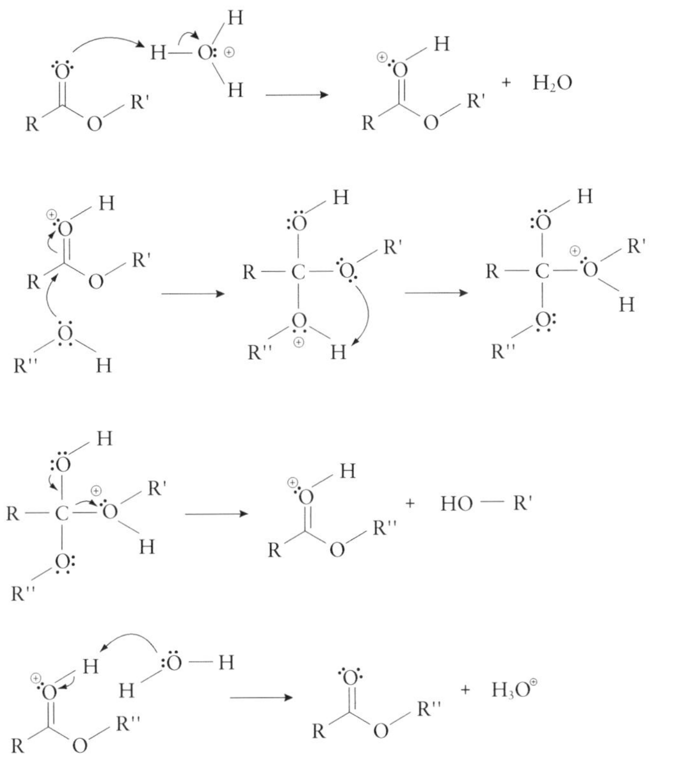

Scheme 2.4 Mechanism of the transesterification reaction in acid conditions

Transamination is induced by primary amines. It is another important reaction that produces the three corresponding amides, a process discussed below in the context of the coupling of furfuryl amine with epoxidised oils.

2.2.2 Reactions of Unsaturated Bonds

Epoxidation, hydroformylation, dimerisation, thiol-ene coupling, oxidative cleavage, and olefin metathesis attack the double bonds of unsaturated oils, fatty acids, or fatty acid esters. These reactions have been exploited for the synthesis of interesting monomers from renewable resources [1–3, 13, 14].

Epoxidation of C=C bonds follows direct or sequential steps, as shown in **Scheme 2.5**, depending on the approach used in terms of catalyst and reaction conditions.

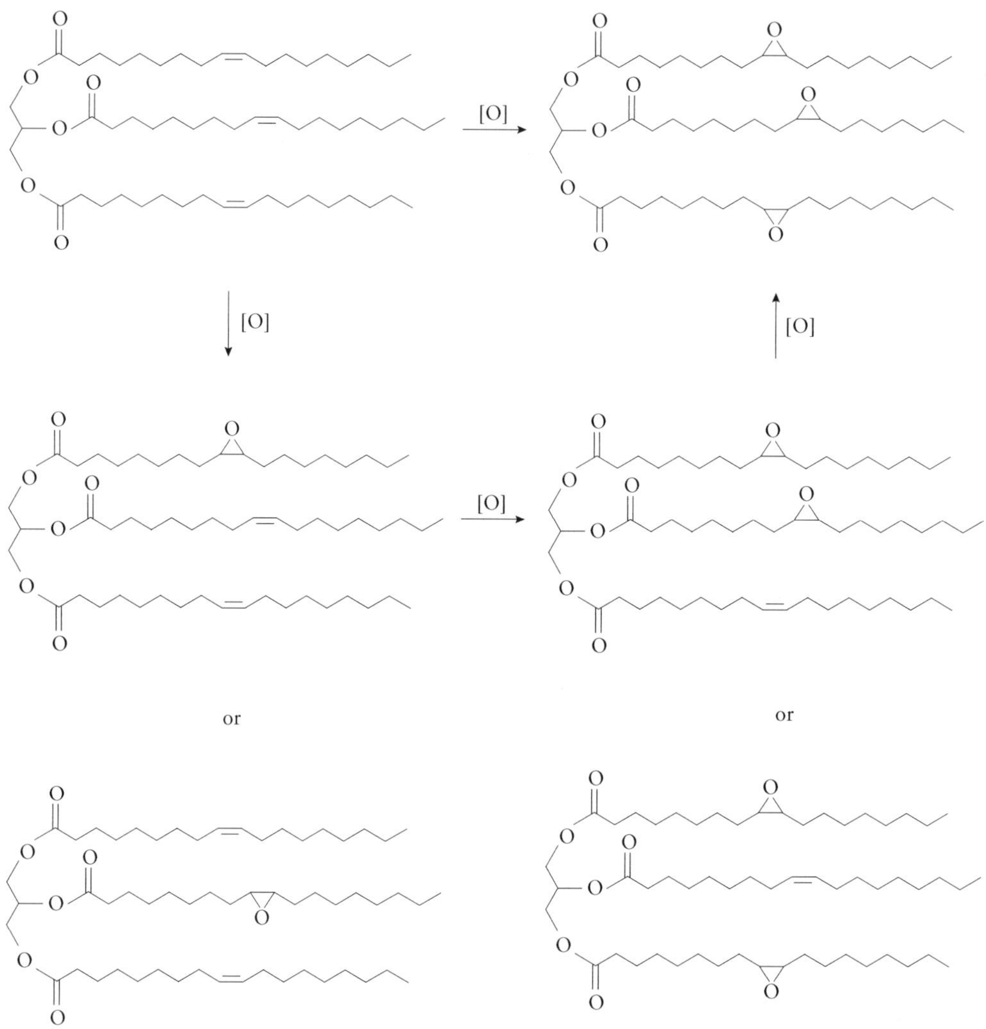

Scheme 2.5 Progressive epoxidation of unsaturated triglycerides applied to trioleate

The hydroformylation reaction converts the unsaturation into a carbonyl moiety, as illustrated in the example of **Scheme 2.6**, in which methyl oleate is converted into methyl-9-stearate with methyl stearate as a byproduct, under the action of a rhodium-based catalyst.

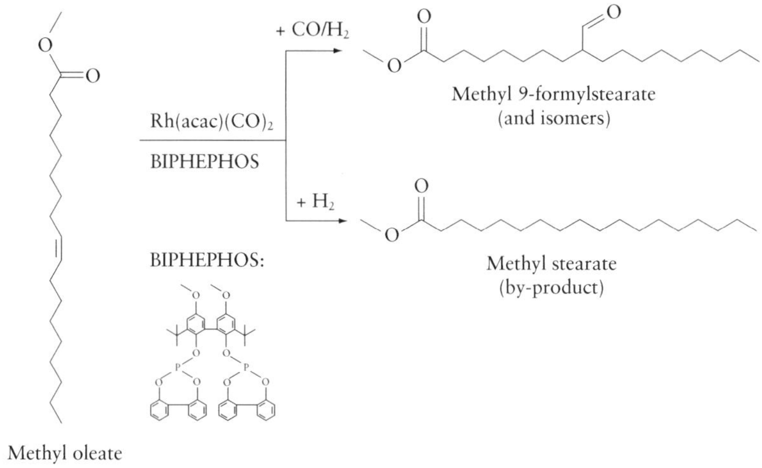

Scheme 2.6 Hydroformylation of methyl oleate

Dimerisation of monounsaturated fatty acids has been exploited to prepare monomers bearing two carboxylic functions and, after reduction, the corresponding diols [2], i.e., potential monomers for the synthesis of polyesters, polyamides, polyethers and polyurethanes incorporating long aliphatic motifs. **Scheme 2.7** shows two possible mechanisms associated with this reaction (i.e., ene- or carbonation coupling), which is inevitably accompanied by further additions to give trimers and oligomers, if not carried out under finely controlled conditions.

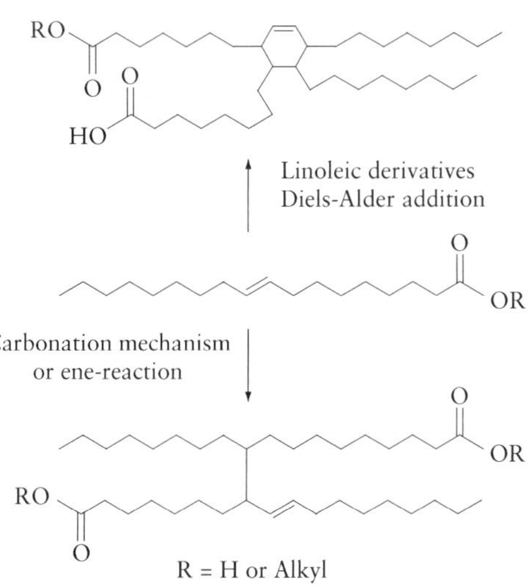

Scheme 2.7 Two possible routes for the dimerisation of oleic moieties [2]

The thiol-ene click reaction has also been applied extensively in recent studies to unsaturated oils and fatty acids. This free-radical process operates on terminal mono-substituted and internal di-substituted unsaturations, the latter coupling being shown schematically in **Scheme 2.8**.

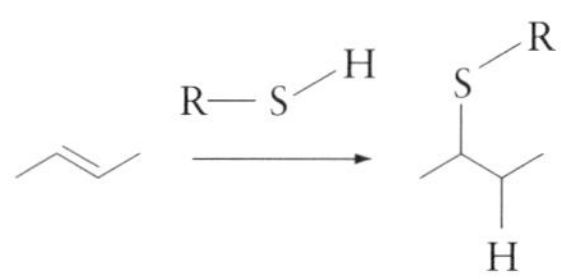

Scheme 2.8 Principle of the thiol-ene reaction

The thiol-ene reaction takes advantage of the easily abstractable hydrogen atom of the thiol group due to the relatively weak sulfur–hydrogen bond [15], which is caused by the low electronegativity of the sulfur atom bonded directly with the electron-deficient hydrogen atom (RSδ– –Hδ+). Cleavage of the S–H bond can be promoted by direct photolysis or indirectly from heat or light that generate nucleophilic alkyl radicals from the initiators. Resulting electrophilic thiyl radicals (RS•) are extremely reactive and can add to a wide variety of unsaturated compounds (electron-rich/poor double-bonds) to form new carbon–carbon linkages. This addition reaction (also termed 'hydrothiolation of C=C bonds') is exothermic [16] and favoured energetically because a new strong σ C–C bond is formed at the expense of a weaker alkene π-bond [17]. However, the overall rate of addition is strongly dependent on the chemical structure of the thiol and ene (internal, isolated, conjugated, non-conjugated, substituted), with thiyl radicals adding faster to electron-rich double-bonds (terminal, mono-substituted) than to electron-deficient ones (e.g., acrylates) [18].

Ozonolysis of triglycerides is another useful reaction that can produce interesting monomers. An example of oxidative scission is given in **Scheme 2.9,** when applied to castor oil in a sequence of modifications leading to polyols and hydroxyacids.

The first step of the ozonolysis mechanism is the initial electrophilic addition of ozone to the C=C double bond to form the molozonide intermediate. Its instability leads to a further reaction, producing a carbonyl and carbonyl oxide molecule (**Scheme 2.10**, II). The carbonyl and carbonyl oxide rearrange to create the stable ozonide intermediate (**Scheme 2.10**, III). A reductive workup is then undertaken to convert the ozonide specie into carbonyl products (**Scheme 2.10**, I) [19].

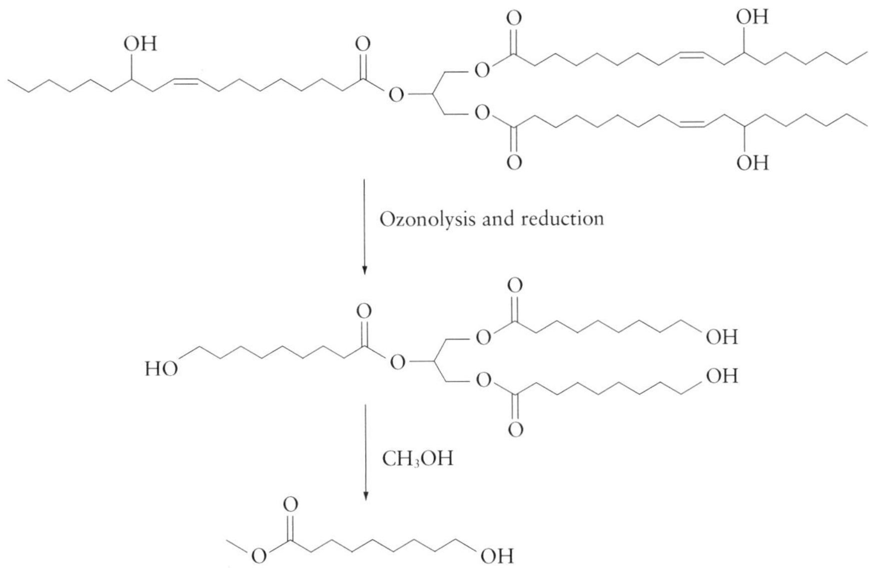

Scheme 2.9 Ozonolysis of castor oil, followed by reduction and hydrolysis

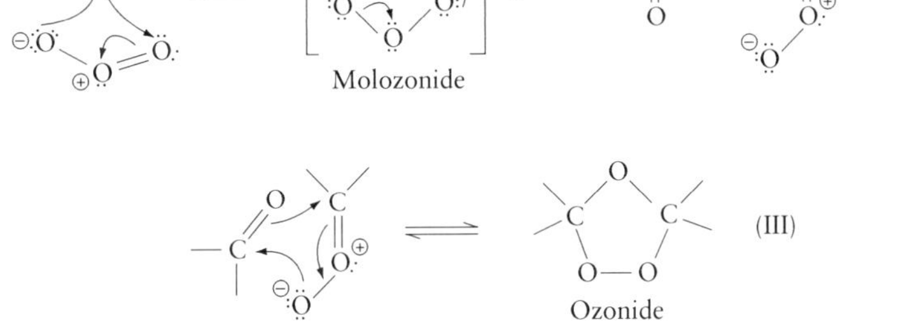

Scheme 2.10 Mechanism of the ozonolysis reaction [19]

The remarkable success of the application of metathesis processes to a vast array of synthetic procedures [20–23] has also transferred into the use of unsaturated oils and their derivatives as substrates [23–33].

In general, olefin metathesis is an organic reaction that allows redistribution of fragments of alkenes by the scission and regeneration of C=C double bonds (**Scheme 2.11**). The metallic catalysts needed for this reaction have evolved rapidly over the past few decades. The impact of this topic in organic synthesis was demonstrated by the 2005 Nobel Prize in Chemistry awarded to Yves Chauvin, Robert H. Grubbs, and Richard R. Schrock for elucidation of the reaction mechanism and discovery of highly efficient and selective metathesis catalysts.

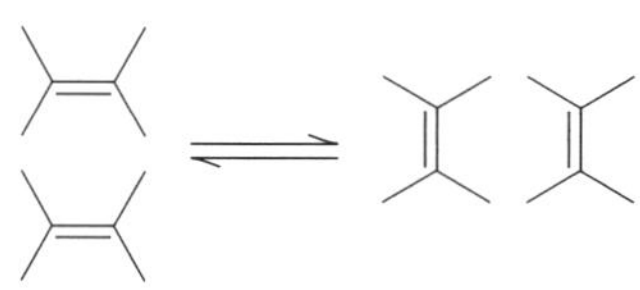

Scheme 2.11 General olefin metathesis reaction

Scheme 2.12 illustrates some applications of this reaction to transformation of castor oil-derived undecenal into aliphatic macrodiols for polycondensation reactions.

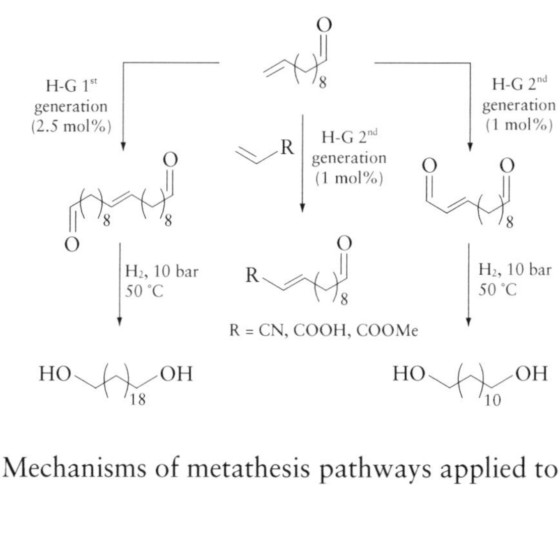

Scheme 2.12 Mechanisms of metathesis pathways applied to undecenal [23]

The succinct outline described above concludes the preliminary notions required for understanding the concepts developed in subsequent chapters, in which the diverse strategies adopted to prepare new macromolecular materials from plant oils are discussed.

References

1. U. Biermann, W. Friedt, S. Lang, W. Lühs, G. Machmüller, J.O. Metzger, M.R. Klaas, H.J. Schäfer and M.P. Schneider, *Angewandte Chemie International Edition*, 2000, **39**, 2206.

2. A. Corma, S. Iborra and A. Velty, *Chemical Reviews*, 2007, **107**, 2411.

3. A. Behr, A. Westfechtel and J.P. Gomes, *Chemical Engineering & Technology*, 2008, **31**, 700.

4. U. Biermann, U. Bornscheuer, M.A.R. Meier, J.O. Metzger and H.J. Schäfer, *Angewandte Chemie International Edition*, 2011, **50**, 3854.

5. S.K. Hoekman, A. Broch, C. Robbins, E. Ceniceros and M. Natarajan, *Renewable and Sustainable Energy Reviews*, 2012, **16**, 143.

6. *Global Biodiesel Market Almost Doubles Every Year between 2001 and 2009*, Renewable Energy Magazine 2010, Global Biodiesel Market Analysis and Forecasts to 2020, Global Data Report 2010.

7. S. Salehpour and M.A. Dubé, *Green Chemistry*, 2008, **10**, 321.

8. S. Salehpour and M.A. Dubé, *Polymer International*, 2008, **57**, 854.

9. S. Salehpour, M.A. Dubé and M. Murphy, *Canadian Journal of Chemical Engineering*, 2009, **87**, 129.

10. A. Behr, J. Eilting, K. Irawadi, J. Leschinski and F. Lindner, *Green Chemistry*, 2008, **10**, 13.

11. F. Yang, M.A. Hanna and R. Sun, *Biotechnology for Biofuels*, 2012, DOI:10.1186/1754-6834-5-13.

12. R. Dobson, V. Gray and K. Rumbold, *Journal of Industrial Microbiology & Biotechnology*, 2012, **39**, 217.

13. M.N. Belgacem and A. Gandini in *Monomers, Polymers and Composites from Renewable Resources*, Elsevier, Amsterdam, The Netherlands, 2008.

14. K. Hill, *Pure and Applied Chemistry*, 2000, **72**, 1255.

15. J. Fossey, D. Leffort and J. Sorba in *Free Radicals in Organic Chemistry*, John Wiley & Sons, New York, NY, USA, 1995.

16. C.E. Hoyle and C.N. Bowman, *Angewandte Chemie International Edition*, 2010, **49**, 1540.

17. A.F. Parsons in *An Introduction to Free Radical Chemistry*, Blackwell Science, Oxford, UK, 2000.

18. M. Claudino in *Thiol-ene Coupling of Renewable Monomers: At the Forefront of Bio-based Polymeric Materials*, Kungliga Tekniska Högskolan, Stockholm, 2011. [Licentiate Thesis]

19. *http://chemwiki.ucdavis.edu/Organic_Chemistry/Hydrocarbons/Alkenes/Reactions_of_Alkenes/Ozonolysis* [Accessed September 2014]

20. *http://www.nobelprize.org/nobel_prizes/chemistry/laureates/2005/* [Accessed September 2014]

21. A. Rybak, P.A. Fokou and M.A.R. Meier, *European Journal of Lipid Science and Technology*, 2008, **110**, 797.

22. H. Mutlu, L. Montero de Espinosa and M.A.R. Meier, *Chemical Society Reviews*, 2011, **40**, 1404.

23. J.C. Mol, *Green Chemistry*, 2002, **4**, 5.

24. J.C. Mol, *Topics in Catalysis*, 2004, **27**, 97.

25. M.A.R. Meier, *Macromolecular Chemistry and Physics*, 2009, **210**, 1073.

26. X. Miao, C. Fischmeister, C. Bruneau and P.H. Dixneuf, *ChemSusChem*, 2009, **2**, 542.

27. T.T.T. Ho, T. Jacobs and M.A.R. Meier, *ChemSusChem*, 2009, **2**, 749.

28. U. Biermann, J.O. Metzger and M.A.R. Meier, *Macromolecular Chemistry and Physics*, 2010, **211**, 854.

29. A. Rybak and M.A.R. Meier, *Green Chemistry*, 2007, **9**, 1356.

30. G.B. Djigoué and M.A.R. Meier, *Applied Catalysis A: General*, 2009, **368**, 158.

31. T. Jacobs, A. Rybak and M.A.R. Meier, *Applied Catalysis A: General*, 2009, **353**, 32.

32. A. Rybak and M.A.R. Meier, *Green Chemistry*, 2008, **10**, 1099.

33. U. Biermann, M.A.R. Meier, W. Butte and J.O. Metzger, *European Journal of Lipid Science and Technology*, 2011, **113**, 39.

3 Polymerisation of Pristine Oils and their Fatty Acids

Activation of the polymerisation of unmodified plant oils and their fatty acids can occur only if unsaturations, –OH groups or naturally occurring epoxy moieties are present in their structure. It follows that triglycerides and their fatty acids bearing straightforward polymethylene chains are not exploitable as potential monomers. Pristine oils incorporating such oxirane-bearing structures (e.g., vernolic acid) are rather exotic and hence unlikely to receive much attention as viable commodities for the synthesis of macromolecular materials. Hence, they are not discussed here and the reader is advised to consult the section below devoted to the much more widespread context of synthetic epoxidised oils. Therefore, this chapter treats exclusively with unsaturated motifs and the unique case of castor oil. Use of both oils and their fatty acids in two major traditional applications [(i) atmospheric siccativity in paints, coatings and lacquers and (ii) alkyd compositions for resins and printing inks] was discussed briefly in **Chapter 1** and is not covered in detail here because it reflects a well-established domain with no relevant recent advances.

3.1 Polymerisation of Unsaturated Oils and Fatty Acids

The C=C unsaturation in all these structures is 1,2-disubstituted, so initiation and propagation by free-radical and cationic mechanisms are sluggish. Hence, they elicit slow processes and poor yields of macromolecular structures though, ultimately, crosslinked materials are generated (as in the case of classical oxido-polymerisation discussed in **Chapter 2**). Preparation of 'blown oils' by thermal treatment >300 ºC is also a poor approach because of substantial losses associated with thermal degradation. It follows that applications are not particularly attractive in practical terms except in the traditional realm of siccative paints and inks.

Use of superacids at moderate temperatures has been the subject of an interesting study [1] applied to oils bearing multiple unsaturations. **Scheme 3.1** shows the mechanism of this reaction, in which the acid-catalysed partial isomerisation of the dienic structure is followed by Diels–Alder coupling with formation of a cyclic adduct. This phenomenon implies that monounsaturated fatty-acid sequences do not react alone but can be exploited if they are mixed with a dienic counterpart, as in the case of oleic and linoleic triglycerides. **Scheme 3.2** shows the mechanism leading to a macromolecular network in the case of an appropriately unsaturated triglyceride. These materials are viscous products or elastomeric gels depending on whether the reaction generates linear (or branched) macromolecules, or networks.

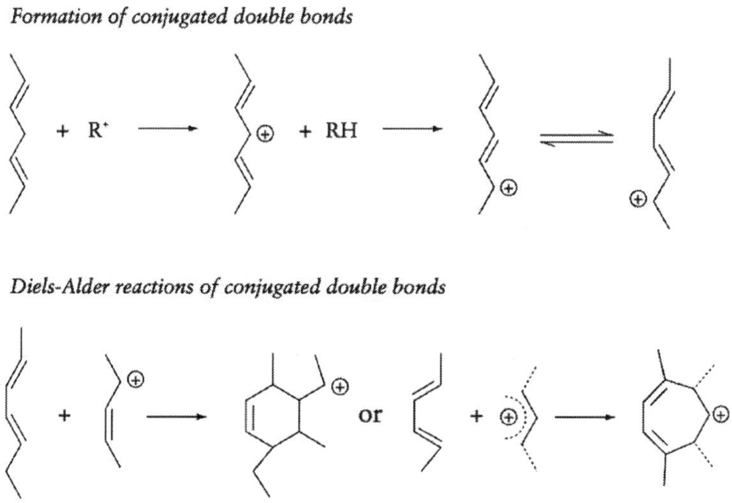

Scheme 3.1 Mechanism of acid-catalysed Diels–Alder coupling between unsaturated oils [1]

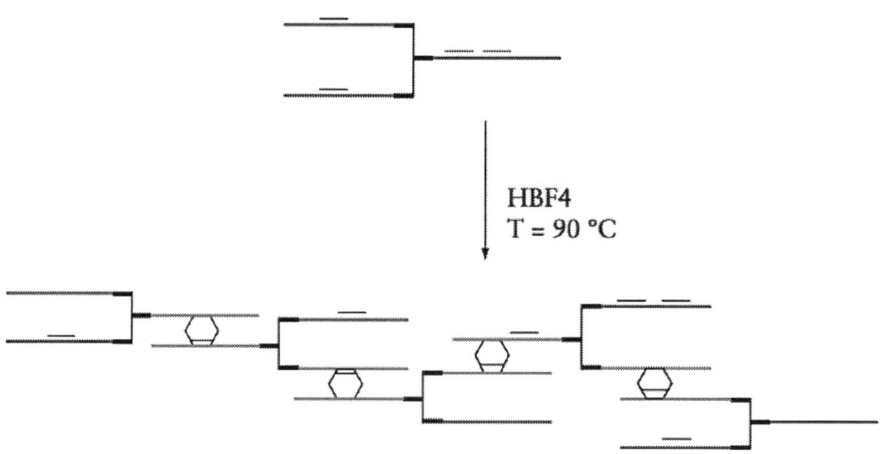

Scheme 3.2 Polymerisation of polyunsaturated structures by superacids [1]

Given that conventional cationic homopolymerisation of unsaturated triglycerides is not viable, the alternative is to use them as comonomers in conjunction with classical reactive monomers such as styrene (ST) or divinylbenzene (DVB) [2–5], as in the example of **Scheme 3.3**. This mechanism resembles the one encountered in the preparation of resins based on unsaturated oligoesters and styrene, in which the readily polymerisable monomer generates growing species that can attack the internal double bond by a grafting process. Oil copolymers prepared by this method are rigid crosslinked materials because of the high glass transition temperature (T_g) associated with polystyrene.

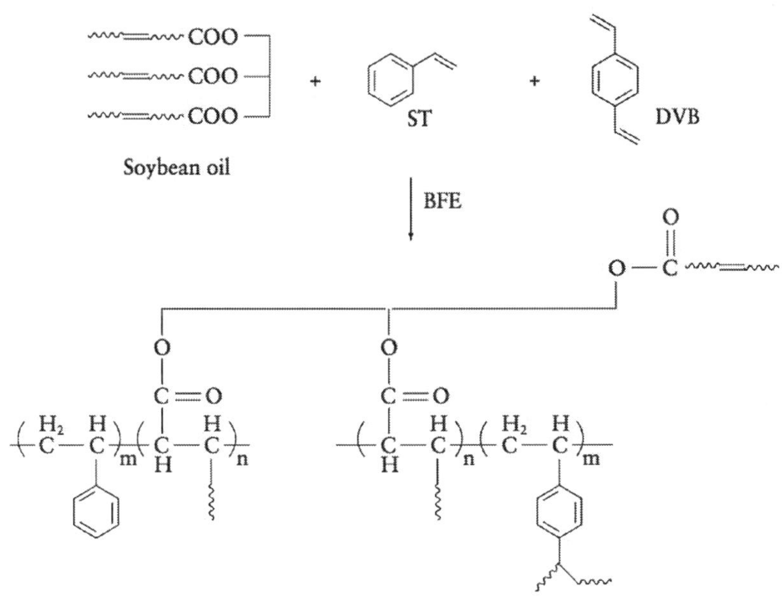

Scheme 3.3 Cationic copolymerisation by boron trifluoride etherate (BFE) of an unsaturated triglyceride and a mixture of styrene and divinylbenzene [2–5]

Use of high pressure through supercritical carbon dioxide (CO_2) has been reported to facilitate cationic oligomerisation of soybean oil with boron trifluoride (BF_3) etherate as a catalyst, giving products suitable as lubricating agents and hydraulic fluids [6].

Use of tung oil as a precursor to macromolecular materials has become an active research topic [7, 8]. Because of its high level of unsaturation, free-radical and cationic polymerisation can be applied to convert this triglyceride into a soft crosslinked material. A recent contribution described the preparation of such networks using BF3 etherate (**Scheme 3.4**) [8]. This study was original because this synthesis was carried out in the presence of the biodegradable polyester polycaprolactone (PCL) to generate semi-interpenetrating networks. Properties of these materials were dependent upon their compositions, with the morphology of a single amorphous phase at 10% PCL, two nanophase features for 20 and 30% PCL contents (two Tg values at about –10 and –50 ºC, with partial crystallisation of the PCL phase) and an interconnected co- continuous microstructure of the two phases if the PCL content was raised to 50%. The storage modulus of these blends in their glassy state was 600–1,800 MPa.

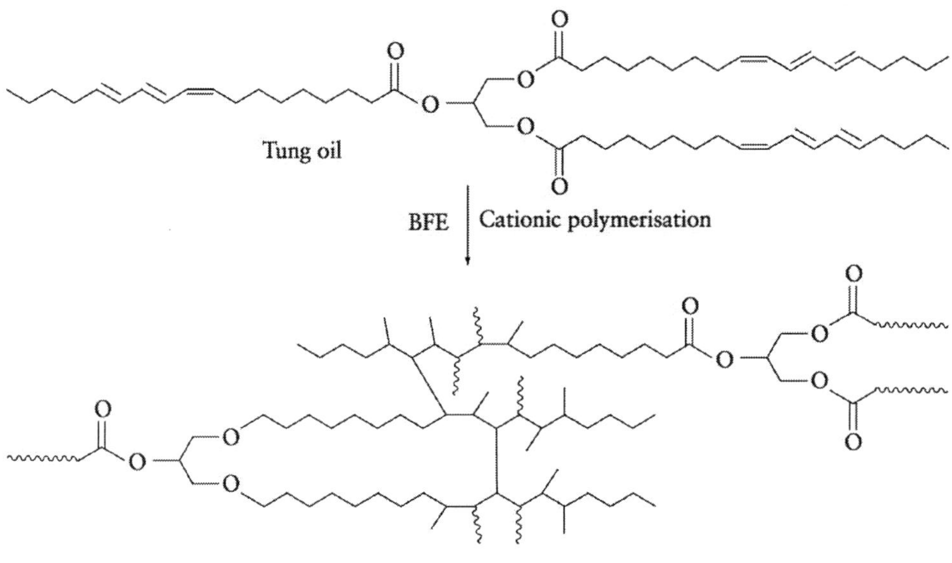

Scheme 3.4 Mechanism for the cationic crosslinking of tung oil [8]

The fact that the three conjugated unsaturations in the tung oil chains behave like a 'diene' moiety in the Diels–Alder reaction has been exploited in original ways to prepare novel polymers. This topic is discussed in **Chapter 7**, which is devoted to application of that 'click chemistry' mechanism.

3.2 Specific Case of Castor Oil

The dominant triglyceride in castor oil (»90%) is that based on ricinoleic acid (**Scheme 3.5**).

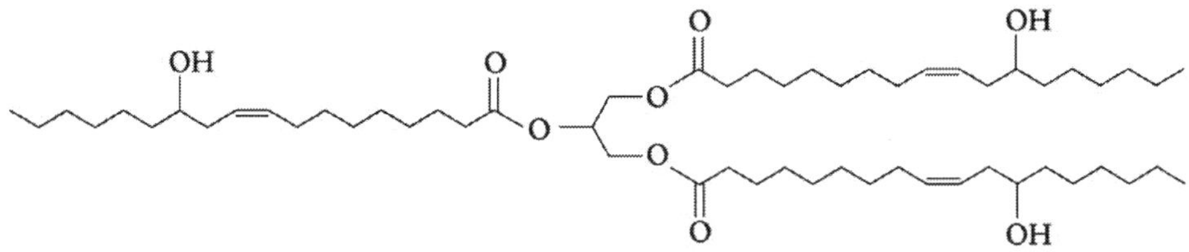

Scheme 3.5 General molecular structure of castor oil

This structural peculiarity, with fatty-acid chains incorporating an unsaturation and a secondary hydroxyl group, opens the potential for direct polymerisation, which also applies to its major ricinoleic fatty acid (albeit in different mechanistic contexts) [11, 12]. The three -OH groups appended on this triglyceride have promptedstudies on the synthesis of polymers such as polyesters and polyurethanes (PU).

It is surprising that relatively little work has been published on the use of *unmodified* castor oils in the synthesis of polyesters compared with studies involving several of its preliminary chemical modifications, as discussed below. A noteworthy excursion into incorporation of castor oil in biodegradable polyesters destined to function as drug carriers [11, 12] describes the melt co-polycondensation of the oil with mannitol, as well as different aliphatic di-and tri-acids. Use of comonomers sebacic acid andcitric acid is shown in **Scheme 3.6**.

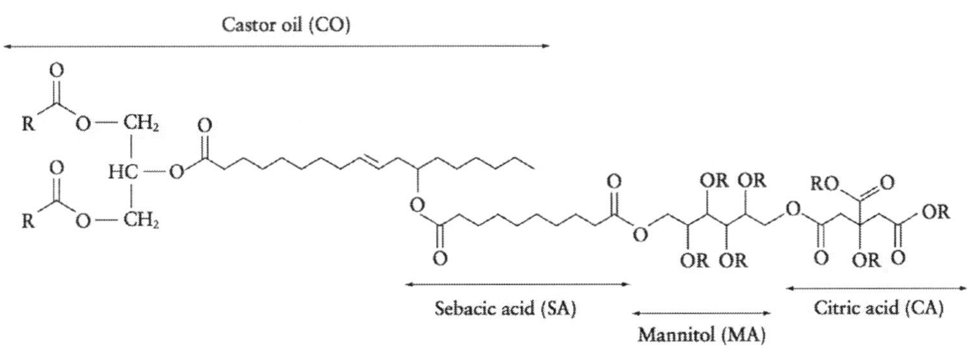

Scheme 3.6 Copolyester based on castor oil and various comonomers [11, 12]

PU based on the use of pristine castor oil were reviewed in 2010 [9, 10], including preparation and properties of an interpenetrating network with other polymers. Recent additions to this topic include: work by Larock and co-workers [13, 14] on waterborne PU and their silica composites; a kinetic study of the reaction of castor oil with 4,4'-methylene diphenyl diisocyanate to produce an elastomeric matrix [15]; the rheology of waterborne PU dispersions for coating applications, prepared with castor oil and isophorone diisocyanate (IPDI) [16] (**Scheme 3.7**); and the synthesis and characterisation of gelled materials arising from the reaction of isocyanate-bearing chitin or chitosan with castor oil [17].

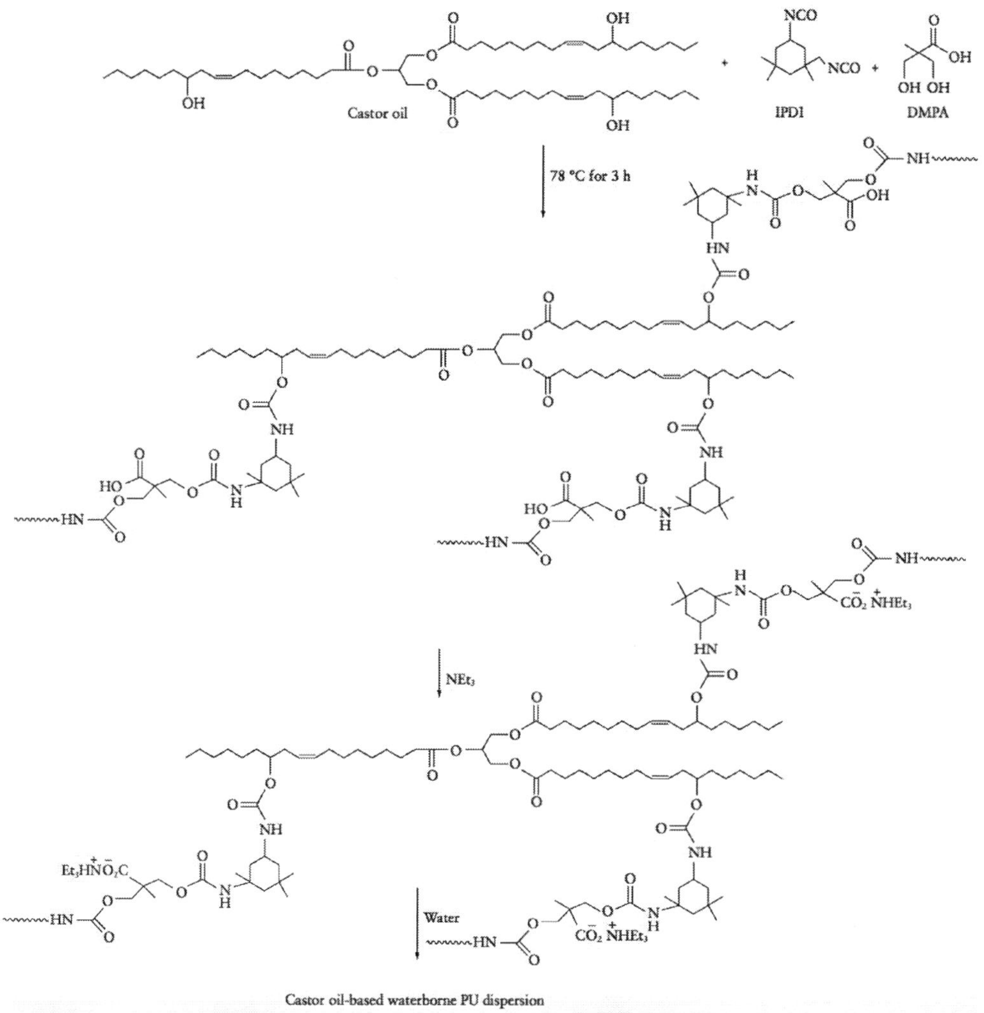

Scheme 3.7 Process leading to a waterborne polyurethane based on castor oil. DMPA: 2,2-Dimethoxy-2-phenylacetophenone [16]

Given the *tris*-OH functionality of castor oil in all the systems mentioned above, the specific stoichiometry associated with these non-linear polycondensations determines whether the ensuing materials will bear a highly branched structure (and hence thermoplastic features) or a networked architecture associated with gelled polymers.

If ricinoleic acid is the actual AB monomer, self-polycondensation leads to the corresponding polyester using conventional and enzymatic polyesterification catalysts [9, 10, 18] (**Scheme 3.8**). Ebata and co-workers [9] studied lipase-based bulk polycondensations applied to the methyl ester so that they could study this polymerisation in detail under green-chemistry conditions. The molecular weight of the ensuing polyesters was 2–100 kDa depending on the enzyme used, and the best results were obtained with lipase from *Pseudomonas cepacia*.

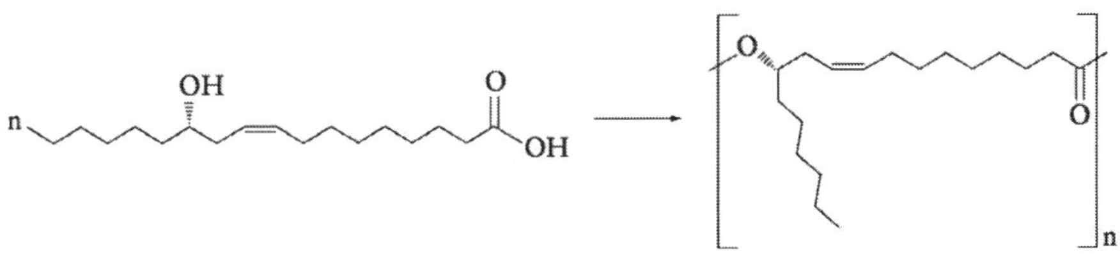

Scheme 3.8 Self-condensation of ricinoleic acid

Aliphatic side chains in each monomer unit are intrinsic structural plasticisers, as evidenced by the very low T_g of the high-molecular-weight polymer (about −75 ºC). After free-radical crosslinking, T_g increased modestly to about −65 ºC showing that, even in a fairly dense network, segmental motions of the dangling aliphatic chains insure high internal macromolecular softness. Further work by the same research team on this polymer focused on its biodegradability by enzymatic hydrolysis [10] to regenerate the acid, which was repolymerised in a cyclic process. Its copolyesterification with diols generates OH–telechelic macromolecules with dangling aliphatic chains, resulting in a low T_g [19] (**Scheme 3.9**).

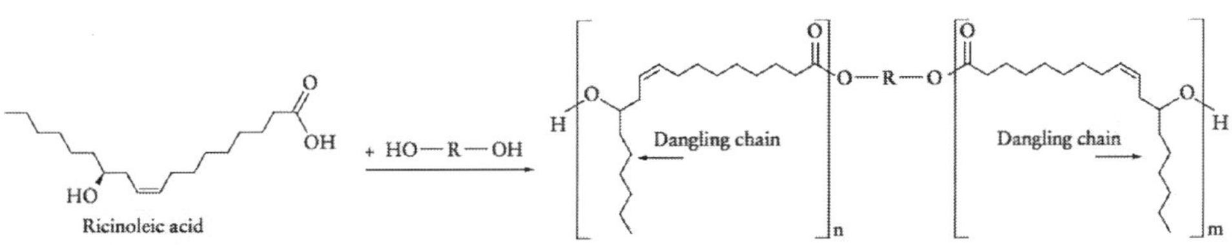

Scheme 3.9 Polycondensation of ricinoleic acid in the presence of diols [19]

If ricinoleic acid is copolymerised with pentaerythritol in bulk using lipases as catalysts [20, 21], star-shaped polyesters are obtained (**Scheme 3.10**). These low-molecular-weight polyol-poly(ricinoleic acid) polyesters possess melting points well below room temperature and high viscosity, suggesting their potential use as biodegradable lubricants and drug-delivering materials.

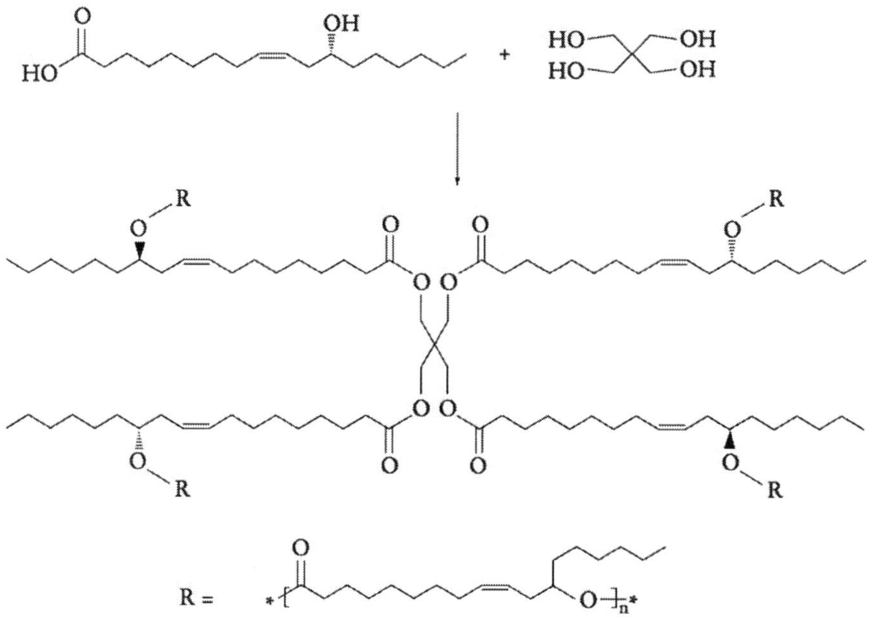

Scheme 3.10 Non-linear polycondensation of ricinoleic acid with pentaerythritol [20, 21]

The research team of Domb has devoted much work on using ricinoleic acid as a comonomer in the synthesis of polyesters for biomedical applications. A recent addition to these wide-ranging investigations deals with polycondensations of ricinoleic acid with a dicarboxylic acid, such as sebacic acid [22]. This reaction produces viscous oligoesters that find applications as biodegradable implants because of their ability to form gels if injected in a buffered medium *in vitro* and *in vivo* at body temperature. Another interesting approach for the synthesis of copolyesters based on ricinoleic acid describes its oligomerisation with propylene glycol and the subsequent use of this macrodiol as an initiator for the ring-opening polymerisation of L-lactide to generate triblock copolymers (**Scheme 3.11**) [23]. Properties of the ensuing materials depend on the block composition, with melting temperatures of

140–180 °C and Young's moduli of 3–1,000 MPa. Thus, potential applications for these ABA coplymers are based entirely on renewable sources.

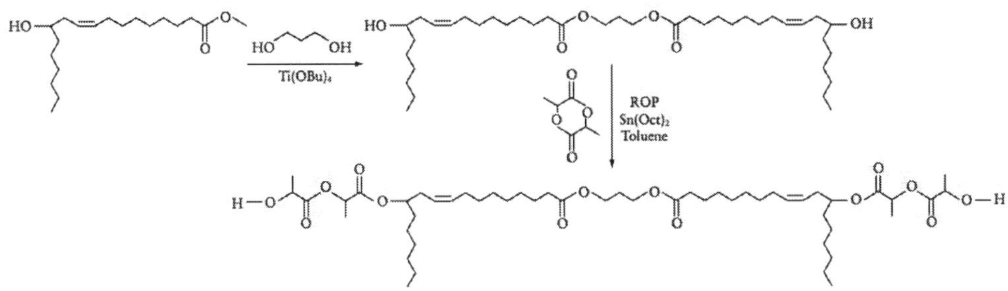

Scheme 3.11 The synthesis of ABA triblock copolyesters based on ricinoleic acid and L-lactide. ROP: Ring-opening polymerisation [23]

A notable recent development has put forward a new approach to the exploitation of plant oils in which polysaccharides are made to react with them or their fatty acids in order to prepare materials entirely based on renewable resources. Thus, starch transesterification with olive and sunflower oil [24] produced a series of soluble macropolyesters with Tg values of 80 to 90 °C and good mechanical properties, which could be cast into transparent films with enhanced surface hydrophobicity. In another vein, soybean oil was coupled with cellulose nanocrystals through thiol–ene chemistry [25] to generate composites which exhibited high tensile strength and maintained a high storage modulus up to 200 °C.

References

1. Z.S. Petrovic, *Contemporary Materials*, 2010, **1**, 39.

2. F. Li, M.V. Hanson and R.C. Larock, *Polymer*, 2001, **42**, 1567.

3. F.K. Li and R.C. Larock, *Journal of Applied Polymer Science*, 2001, **80**, 658.

4. F.K. Li and R.C. Larock, *Journal of Applied Polymer Science*, 2002, **84**, 1533.

5. F.K. Li, J. Hasjim and R.C. Larock, *Journal of Applied Polymer Science*, 2003, **90**, 1830.

6. L. Zengshe, S. Brajendra and S. Erhan, *Biomacromolecules*, 2007, **8**, 233.

7. T.M. Lacerda, A.J.F. Carvalho and A. Gandini, *RSC Advances*, 2014,**4**, 26829.

8. S.A. Madbouly, K. Liu, Y. Xia and M.R. Kessler, *RSC Advances*, 2014, **4**, 6710.

9. H. Ebata, K. Toshima and S. Matsumura, *Macromolecular Bioscience*, 2007, **7**, 798.

10. H. Ebata, K. Toshima, and S. Matsumura, *Journal of Synthetic Organic Chemistry - Japan*, 2008, **66**, 673.

11. P.S. Sathiskumar and G. Madras, *Polymer Degradation and Stability*, 2011, **96**, 1695.

12. P.S. Sathiskumar, S. Chopra and G. Madras, *Current Science - India*, 2012, **102**, 97.

13. Y. Xia and R.C. Larock, *Macromolecular Materials and* Engineering, 2011, **296**, 703.

14. Y. Xia and R.C. Larock, *Macromolecular Rapid Communications*, 2011, **32**, 1331.

15. A.C. Milanese, M.O.H. Cioffi, H.J C. Voorwald and C.Y. Shigue, *Journal of Applied Polymer Science*, 2011, **122**, 3168.

16. S. A. Madbouly, Y. Xia and M.R. Kessler, *Macromolecules*, 2013, **46**, 4606.

17. R. Gallego, J.F. Arteaga, C. Valencia and J.M. Franco, *Molecules*, 2013, **18**, 6532.

18. D.S. Ogunniyi, *Bioresource Technology*, 2006, **97**, 1086.

19. Z.S. Petrovic, I. Cvetkovic, D. Hong, X. Wan, W. Zhang, T. Abraham and J. Malsam, *Journal of Applied Polymer Science*, 2008, **108**, 1184.

20. A.R. Kelly and D.G. Hayes, *Journal of Applied Polymer Science*, 2006, **101**, 1646.

21. D.G. Hayes, *ACS Symposium Series*, 2006, **939**, 126.

22. A. Shiranov and A. J. Domb, *Biomacromolecules*, 2008, **7**, 288.

23. T. Lebarbe, E. Ibarboure, B. Gadenne, C. Alfos and H. Cramail, *Polymer Chemistry*, 2013, **4**, 3357.

24. Z. Soyler, M.A.R. Meier, *ChemSusChem*, 2017, **10**, 182.

25. L. Song, Z. Wang, M. E. Lamm, L. Yuan, C. Tang, *Macromolecules*, 2017, **50**, 7475.

4 Monomers and Polymers from Chemically Modified Plant Oils and their Fatty Acids

Most contributions to the preparation of macromolecular materials derived from vegetable oils involve at least a modicum of chemical transformation, if not major modifications. The paucity of studies associated with their direct exploitation is due to the intrinsic mechanistic limitations of these pristine structures (with the possible exception of castor oil and ricinoleic acid) in terms of constructing as wide a variety of materials as required by the vast array of polymer applications.

This section attempts to cover all the relevant topics related to the synthesis of modified oil derivatives, their polymerisation, and the properties of the ensuing materials. The reader is referred to **Chapter 2** for basic considerations concerning the mechanisms of these modifications.

4.1 Epoxidised Structures

Conversion of internal unsaturations in triglycerides and their fatty acids into epoxide moieties has been optimised to enable their industrial implementation an economically viable operation with some oils. This strategy has opened the way to elaboration of macromolecular materials that could become real commodities. This *rationale* implies that the performance of these polymers should adequately match the respective applications they are conceived to fulfill. Apart from such common epoxidised oils, this section deals with other more elaborate monomers in which the oxirane function is present as an end group and hence more reactive.

4.1.1 Direct Polymerisation

Photo-initiated cationic polymerisation of epoxidised oils has been the subject of intense scrutiny by the research team led by Crivello, starting with the groundbreaking study in 1992 [1–3] applied to several epoxidised triglycerides, with the linseed homologue shown in **Scheme 4.1**. Since that contribution, this [1–3] and other research teams have carried out further research, including use of thermally activated cationic initiators and incorporation of fillers to prepare composites with improved mechanical properties [4].

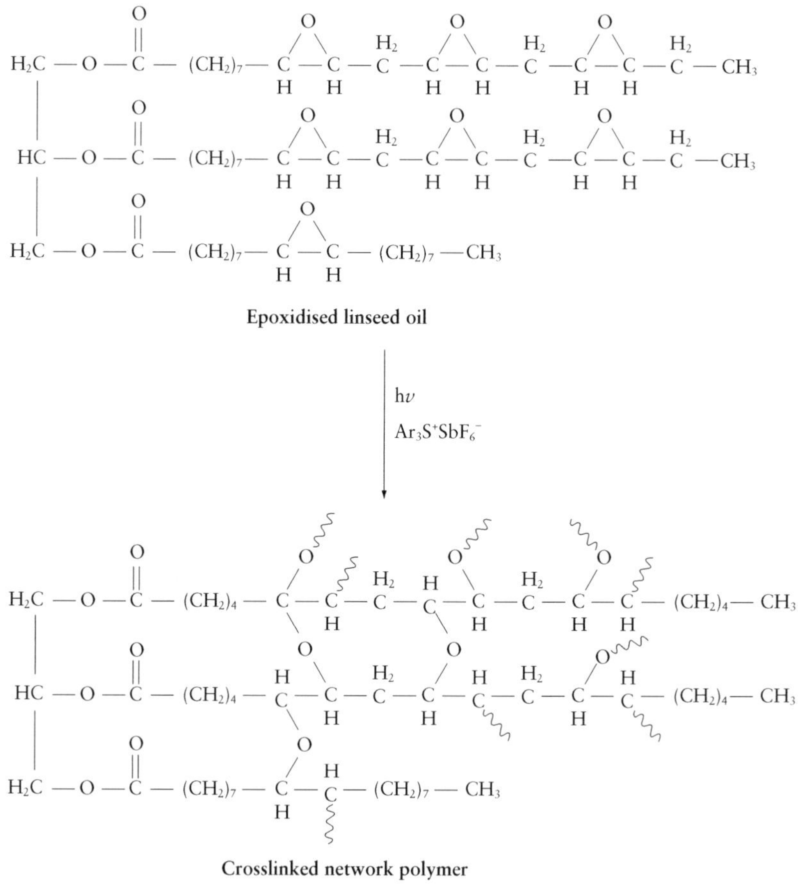

Scheme 4.1 Cationic photocrosslinking of epoxidised linseed oil [1–3]

Latent initiators that are activated photochemically or by heating, such as benzylpyrazinium salts, enable controlled polymerisations. In fact, the activity of pyrazinium salts can be controlled by electronic modifications of benzyl and pyrazine groups [5]. Epoxidised soybean and castor oils have been polymerised by cationic means in the presence of the latent initiator *N*-benzylpyrazinium hexafluoroantimonate [6]. Cationic resins prepared from epoxidised castor oil display higher glass transition temperature (T_g) values and lower coefficients of thermal expansion than counterparts prepared from epoxidised soybean oil (ESO) because of increased intermolecular interactions in the former materials when compared with the latter polymer [6].

Both epoxidised oils have also been copolymerised with the diglycidyl ether of bisphenol A to give networks with better mechanical properties [7, 8]. For example, a copolymer containing 10 wt% of epoxidised castor oil produces an epoxy resin with better interfacial mechanical properties, and addition of a large amount of soft segments results in a decrease in the crosslink density and an increase in the toughness of the final copolymer.

An original alternative to these approaches involving epoxidised triglycerides relates to the bulk cationic polymerisation of epoxidised methyl oleate (i.e., a monofunctional monomer) by $HSbF_6$ at room temperature (RT) [9]. The resulting structure was, therefore, a linear polyether macrodiol whose potential as a materials precursor stemmed from the possibility of converting it to a macropolyol for the synthesis of oil-based polyurethanes (PU) by reduction of some of the ester groups present in each monomer unit using lithium aluminium hydride (LAH) [9]. The sequence of these operations is given in **Scheme 4.2**. The same monomer was also polymerised using an ionic coordination initiator [10].

Cationic polymerisation of epoxidised oils by boron trifluoride (BF_3) has also been conducted in liquid carbon dioxide (CO_2) [11].

Fully biobased epoxidized plant oil thermosets were recently synthesized by reacting epoxidized soybean oil with dicarboxylic acid oligomers derived from castor oil (**Scheme 4.2**) through a catalyst-free curing method [12].

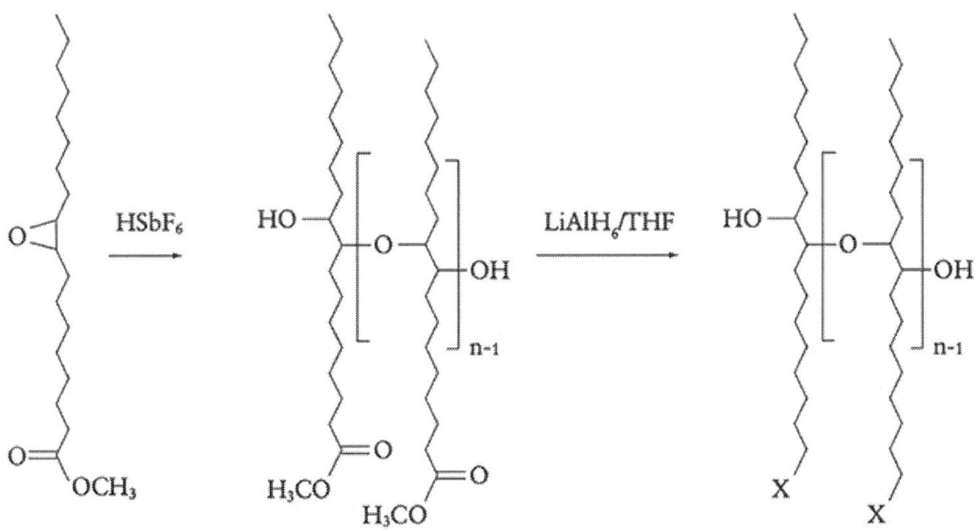

Scheme 4.2 Cationic polymerisation of epoxidised methyl oleate and subsequent partial reduction of ester groups of the product into primary OH groups ($X=CH_2OH$, or $COOCH_3$). THF: Tetrahydrofuran [9]

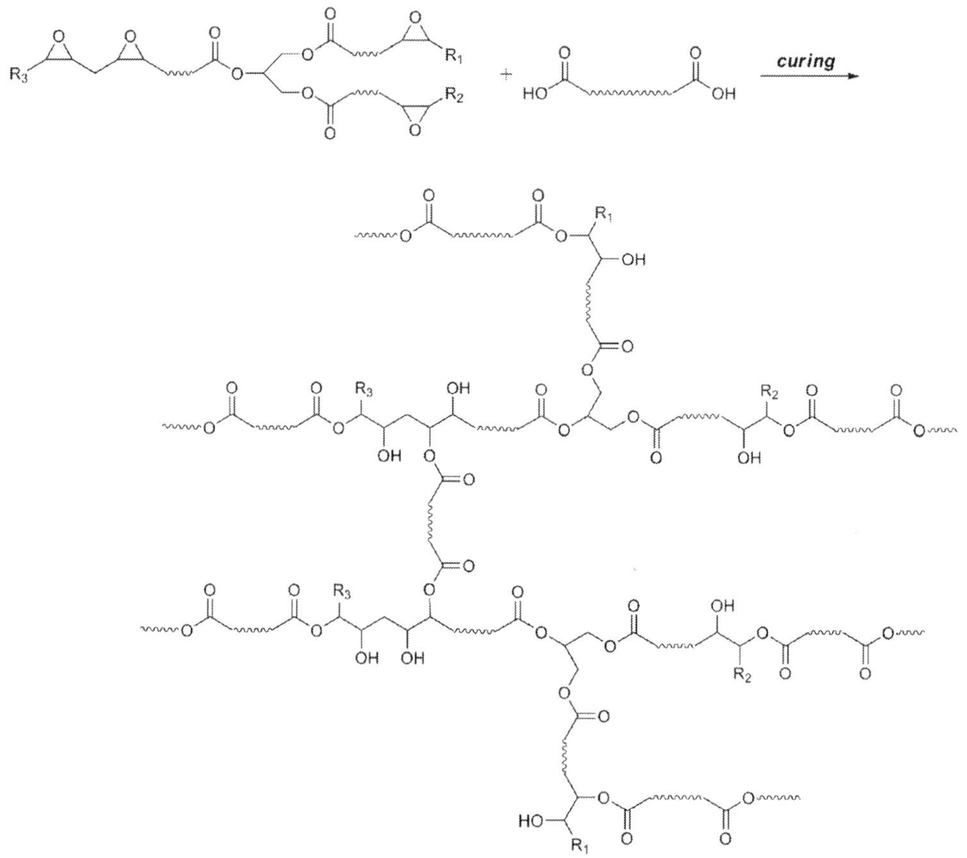

Scheme 4.3 Epoxidized soybean oil for the preparation of epoxy thermoset [12].

4.1.2 Reactions with Amines and Anhydrides

Classical combination of diepoxide monomers with diamine counterparts in conventional Araldite-type adhesives involves terminal oxirane groups, which are considerably more reactive than the internal counterparts present in epoxidised oils or fatty acids. This difference is probably why studies on the potential of coupling epoxidised oils and diamine for the development of epoxy resins based on renewable resources have been published.

Téllez and co-workers [13] undertook reaction of epoxidised linseed oil (ELO) with di- and tri-amines at 30–70 °C to prepare aminated derivatives (AELO), like those shown in **Scheme 4.4**, which were characterised and then crosslinked at 180–220 °C. The thermoset prepared with *p*-xylylenediamine was a very rigid material due to incorporation of aromatic rings and which exhibited the highest decomposition temperature. However, preparation of this thermoset required the highest curing temperature and longest reaction time. Regrettably, other properties of the ensuing materials were left for a subsequent study, which has yet to be published.

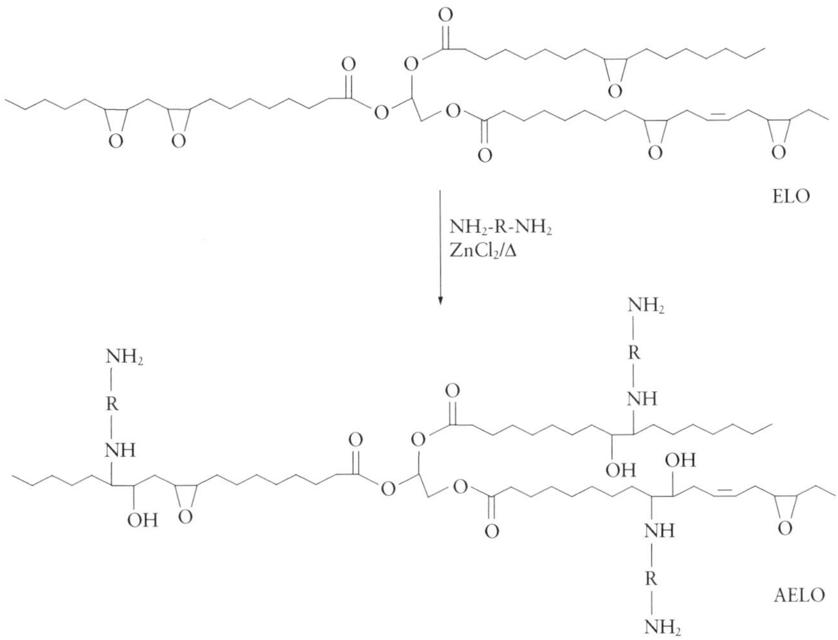

Scheme 4.4 Reaction of epoxidised linseed oil with diamines [13]

Czub [14, 15] carried out a thorough investigation on the use of various epoxidised oils in conjunction with bisphenol A in the preparation of epoxy networks with different hardeners (including dicyanodiamide and triethylenetetramine) and followed the curing reactions. However, the ensuing materials were not characterised.

Different amines and different ratios of the amine to the epoxy have been used to cure several epoxidised vegetable oils [16]. The bis(*p*-aminocyclohexyl)methane diamine afforded the highest performance and amine:epoxy ratios of 0.8 and 1.0 gave thermosets with superior performances.

The reaction of ESO with decamethylene diamine has been examined using a wide range of reagent stoichiometries [17]. When an excess of diamine was employed, aminolysis of the triglyceride produced fatty-acid residues with two oxirane moieties and so non-crosslinked elastomers with a low T_g (–30 to –17 °C) were produced. When the molar excess of diamine was twofold, higher molecular weights were obtained and viable bioelastomers were obtained by their reaction with succinic anhydride. These bioelatomers possessed low damping properties and adequate mechanical properties for engineering applications.

Epoxidised di- and tri-10-undecenoyl macromonomers have been the subject of an innovative investigation that allowed terminal epoxy groups to be inserted into fatty-acid structures, including some [18] incorporating phosphorous-containing moieties (**Scheme 4.5**). These novel monomers were polymerised through the standard epoxy process using aromatic diamines because the oxirane moiety was now mono-substituted and hence displayed a much higher reactivity than their internal counterparts. Introduction of phosphor-containing moieties was aimed at reducing the combustibility of the final materials.

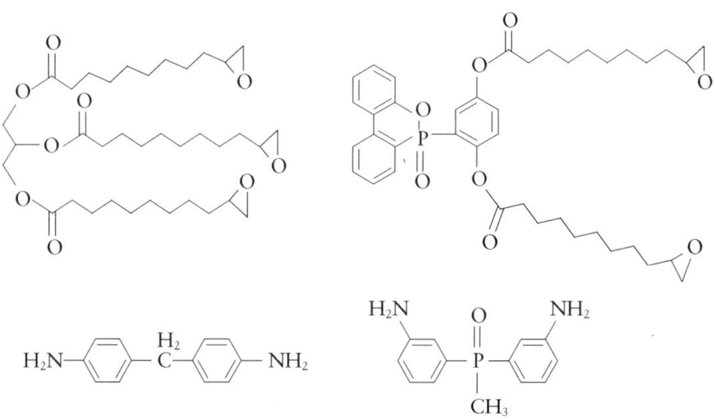

Scheme 4.5 Monomers used to prepare epoxy resins based on terminally epoxidised derivatives of fatty acids [18]

A more recent study of the reaction of primary amines with epoxidised vegetable oils is discussed in **Chapter 7**. It deals with the Diels–Alder reactions involving furan derivatives.

Use of anhydride hardeners to crosslink epoxidised oils has received more attention. Boquillon and Fringant [19] studied the kinetics of the polycondensation of epoxidised linseed oil with various aromatic and cycloaliphatic anhydrides at 150–170 °C and characterised the ensuing networks by measuring their T_g and flexural modulus (which both increased with increasing proportion of anhydride) as well as dynamic mechanical properties and their crosslink density. These materials, based partly on renewable resources, displayed adequate properties in terms of applications of thermosets associated with a relatively low T_g (35–110 °C). A similar investigation in which epoxidised vegetable oils substituted with ≤50% of bisphenol A diglycidyl ether gave networks with a high T_g and high elastic moduli, together with improved impact strength and fracture toughness, compared with the counterpart without the epoxidised oil [20]. Nanocomposites consisting of these matrices and various inorganic fillers were also studied [21].

The reaction of ESO with a terpene-based anhydride and the properties of the ensuing network were compared with those of its polymerisation with hexahydrophthalic anhydride [22]. The material based entirely on renewable resources had a higher T_g and stronger mechanical features that were improved further by adding Lyocell cellulose fibres as a reinforcement agent. Cellular morphologies were prepared by the reaction of ESO with malonic acid because the malonic monoester intermediate decarboxylates at ≈135 °C, thus generating an intrinsic CO_2 blowing agent [23]. These foams were characterised in terms of density, compressive modulus, rebounding of cell morphology, and biodegradability. They showed viable properties, including a high rate of biodegradation.

A study on the preparation of thermosetting resins was reported in which ESO was used in variable partial replacement of the synthetic epoxy prepolymer based on the diglycidyl ether of bisphenol A, with methyltetrahydrophthalic anhydride as a crosslinking agent and 1-methyl imidazole as an initiator [24]. Effects of the replacement of increasing amounts of the oil derivative on network properties, such as the storage modulus in glassy and rubbery regions, T_g, and impact and compressive properties, were examined. The formulation containing 40 wt% of epoxidised oil resulted in a resin with an optimum set of properties, elastic modulus in the glassy state was 93% that of the neat reference resin; T_g decreased only by ≈10 °C; impact strength increased by ≈38%, without loss of transparency.

4.1.3 Acrylation Reactions

The reaction of epoxidised vegetable oils with acrylic acid or its anhydride has been studied thoroughly. As an alternative to counterparts based on fossil resources, triglycerides bearing multiple acrylic functions represent a particularly interesting family of macromonomers for the development of photocurable resins from renewable resources. Acrylated epoxidised soybean oil (AESO) synthesised from the reaction of acrylic acid with ESO [25, 26] is available under the brand name Ebecryl 860 from UCB Chemicals Company. AESO has been studied extensively in polymers and composites, particularly as a photosensitive coating material [25, 26]. The reaction takes place in two steps with very different kinetic features [25, 26]. The ring-opening is rather fast, whereas esterification of the resulting secondary OH group requires much more time to reach completion because of steric hindrance and its different mechanistic features (**Scheme 4.6**). Free-radical photo-crosslinking of a fully acrylated soybean oil film with ≈8 acrylic moieties per triglyceride can be achieved in a few seconds [25, 26], making the process very attractive for continuous coating of packaging substrates.

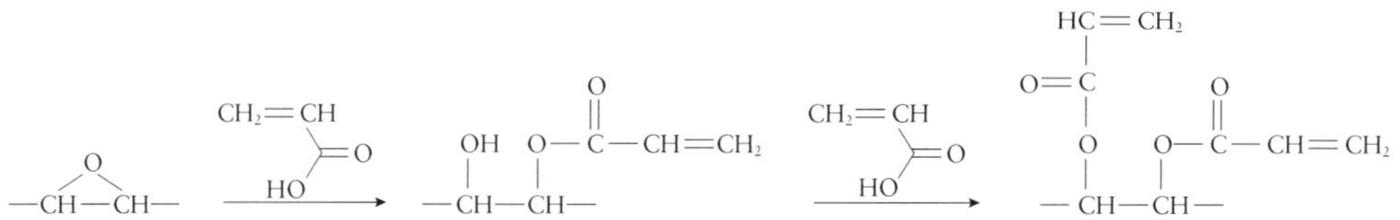

Scheme 4.6 Two-step acrylation of an oxirane moiety incorporated into a triglyceride chain [25, 26]

Acrylated oils can also be blended with reactive diluents such as styrene to improve processability and afford suitable thermosets and composites for structural applications [27, 28] by changing the acrylation level of the triglyceride and by varying the proportion of styrene.

Work has also been done to further modify partially acrylated vegetable oils incorporating hydroxyl groups, in particular by their reaction with maleic anhydride (**Scheme 4.7**). Its copolymerisation with styrene resulted in thermosets with higher crosslink densities, T_g and storage moduli than corresponding materials involving the acrylated precursor [29].

A very interesting and original contribution to the realm of epoxy resins derived from plant oils was published recently in which new monomers bearing terminal oxirane functions were synthesised from soybean and linseed oil [30]. The synthetic pathway is shown in **Scheme 4.8**, in which only the oleic structure of soybean is portrayed, although the mechanistic steps apply to all unsaturated chains in the oils. After standard hydrolysis and re-acidification leading to a mixture of the corresponding free fatty acids (FFA), unsaturated structures are separated and their terminal carboxylic group converted into a glycidyl function by reacting it with epichlorohydrin (EPCH). The ensuing terminal oxirane fatty acids are finally submitted to a well-known epoxidation procedure using *meta*-chloroperoxybenzoic acid (MCPBA), applied to their internal unsaturations, to give novel monomers incorporating two or three oxirane rings, of which one sits as an end-group.

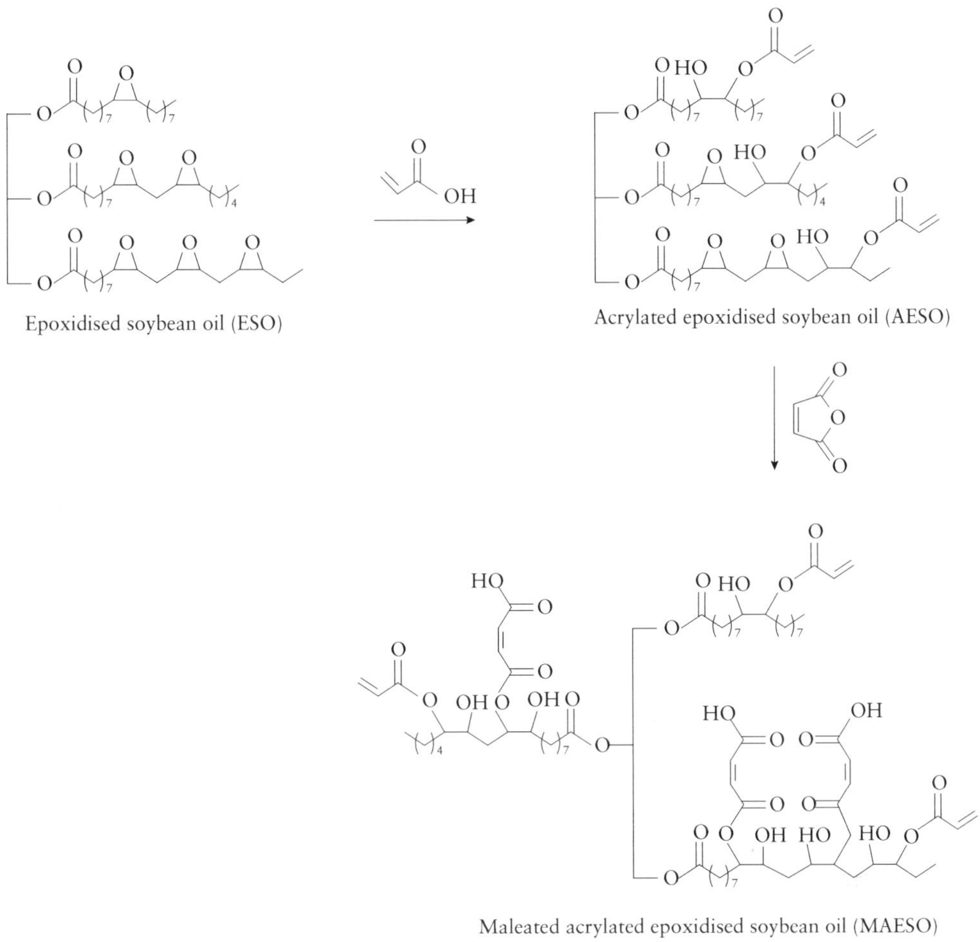

Scheme 4.7 Partially acrylated epoxidised soybean oil followed by its maleation [29]

The authors then proceeded to conduct a thorough study of the preparation of a large selection of epoxy resins using these new monomers together with several epoxy comonomers, including conventional ESO and the classical fossil-derived aromatic bisphenol A diglycidyl ether (DGEBA), and treating the mixture with the 4-methyl-1,2-cyclohexanedicarboxylic anhydride (MHHPA) curing agent (**Scheme 4.9**). Single epoxy monomers were also polymerised by cationic means using a boron fluoride monoethyamine complex [30].

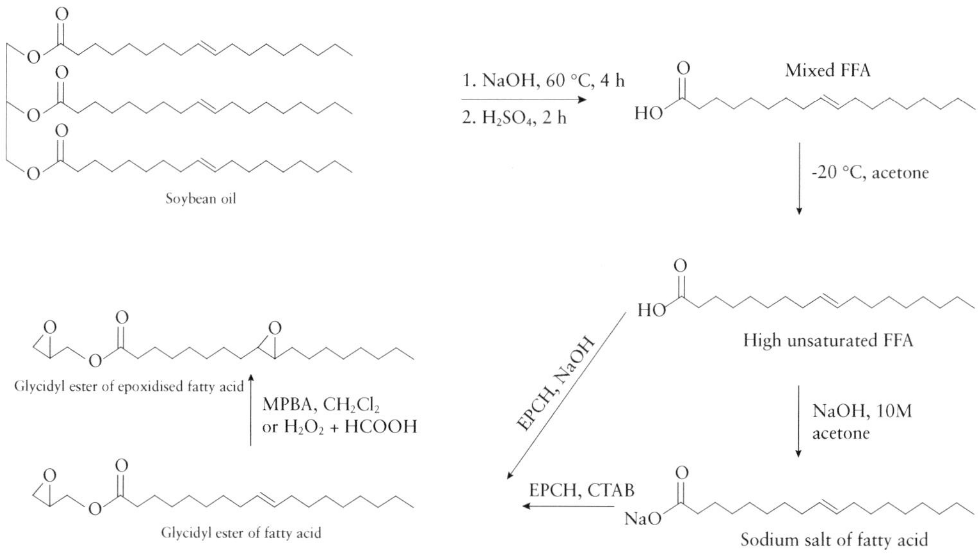

Scheme 4.8 Synthesis of novel epoxydised fatty acids from unsaturated vegetable oils as exemplified by the specific case of soybean oil. CTAB: Cetyltrimethylammonium bromide [30]

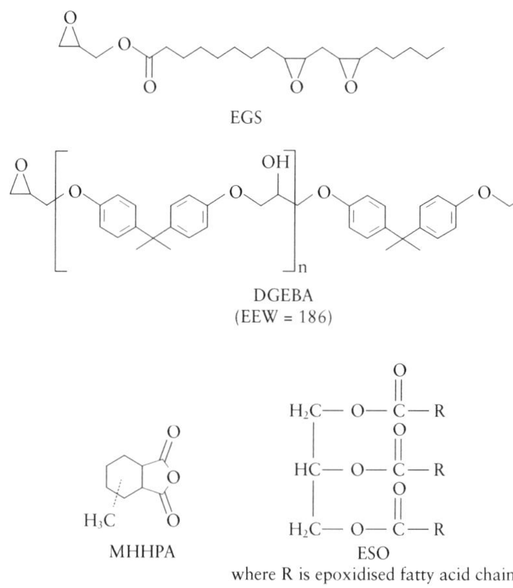

Scheme 4.9 Monomers used in the study of new epoxy resins. EGS: Glycidyl ester of epoxidised soybean oil fatty acid and EEW: epoxy equivalent weight [30]

Results of this methodical investigation [30] provided a comprehensive view of how the polycondensations proceeded as a function of the composition of a given crosslinking system and produced a set of key parameters (e.g., T_g, thermal degradation, network density, tensile strength, flexural strength, modulus) and the role of the viscosity of the initial monomer mixture on processability (use of reactive diluents). This work shows how the search for optimised compositions based on renewable resources can contribute substantially to the scientific and technological knowledge of epoxy resins derived from vegetable oils.

4.2 Polyol Structures for Polyurethanes

Incorporation of hydroxyl groups in triglycerides and their fatty acids constitutes a major research area, particularly within the broad context of the synthesis of biological-based PU. C=C unsaturations in plant oils can be modified by different chemical mechanisms to introduce –OH moieties (**Scheme 4.8**) [31]. One of the most representative approaches is opening the oxirane ring in epoxidised oils by various reagents (**Scheme 4.10**).

Polyols from sunflower, canola, soybean, sunflower, corn, and linseed oils have been prepared from epoxidised plant oils by their reaction with methanol [32]. Their polycondensations with diisocyanates could generate PU networks readily. Polyols from linseed oil gave the highest crosslink densities and best mechanical properties, whereas those from sunflower oil gave soft materials with the lowest T_g and strength. Properties of the ensuing PU prepared with a given diisocyanate were dependent mainly on their crosslink density and less from the position of the reactive sites in the fatty-acid chains [32]. That is, the higher the degree of unsaturation of the starting oil, the higher the network density attained.

Use of hydrogenation, *viz.* another mechanistic approach to convert oxiranes into OH groups, was applied to ESO [33]. The ensuing polyols displayed OH numbers ranging from 80 to 225 depending on the reaction time. Those with high OH content gave glassy PU, whereas those with modest hydroxylation produced elastomeric materials. The same epoxidised oil was converted into a polyol based entirely on renewable resources by the reaction with lactic acid [34]. This polyol bore two OH functions per converted oxirane moiety. Hence, it generated PU with a peculiar network morphology that enabled preparation of materials with 'tunable' T_g over a wide range.

PU foams based on vegetable oils have also been investigated. Variably epoxidised rapeseed oil and subsequent oxirane ring-opening with diethylene glycol produced two polyols with 2.5 and 5 OH groups per triglyceride, respectively [35]. Foams prepared with the former displayed higher resilience and elongation at break, whereas those based on the latter exhibited higher tensile and compression strength, as well as a better cell morphology. Partial substitution of a fossil-based polyether polyol with one prepared from epoxidised palm oil using hexamethylene glycol gave flexible foams with improved homogeneity in terms of cell size and improved compressive strength [36].

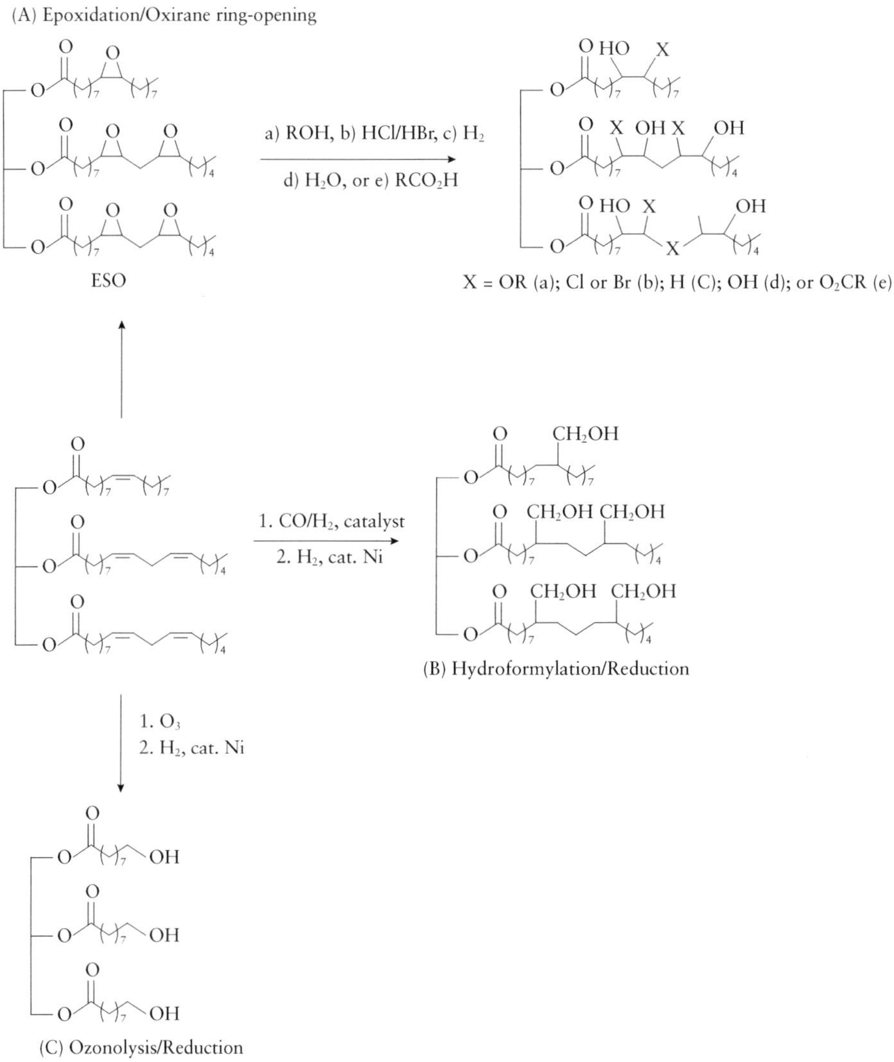

Scheme 4.10 Different mechanistic pathways leading to incorporation of OH groups in triglyceride chains through their unsaturations [31]

An original alternative way to prepare polyols was applied to ESO that was treated first with CO_2 to give the corresponding carbonate, and which was then reacted with ethanolamine to generate OH groups (**Scheme 4.11**) [37, 38]. The reaction of these polyols with blocked polyisocyanates produced PU networks with remarkable thermal properties and electrical-insulation aptitudes. Blends of the polyols with polypropylene glycol widened the tenability of the physical properties of the ensuing PU [38].

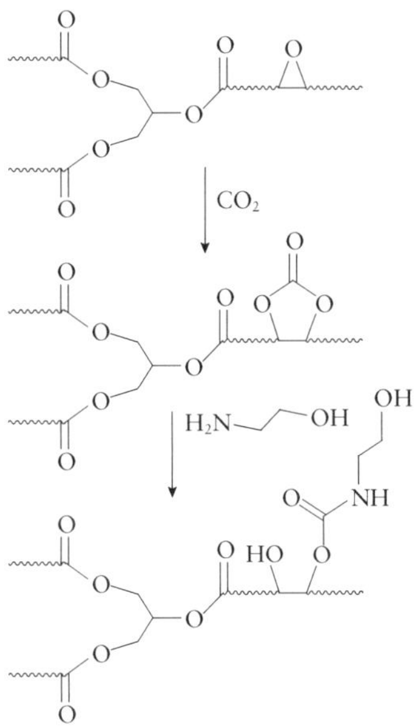

Scheme 4.11 Mechanism of conversion of an oxirane moiety into a primary OH group through carbonatation [37]

The ozonolysis and subsequent hydrogenation (**Scheme 4.10**) of epoxidised canola oil produced polyols that were used to prepare fully bio-based PU by coupling them with a diisocyanate derived from a fatty acid [39, 40]. Properties of these novel PU were comparable with those of counterparts prepared with the same polyols and the conventional fossil-based 1,6-hexamethylene diisocyanate.

Lu and Larock [41–44] carried out extensive work on the preparation of waterborne PU based on soybean oil polyols and isophorone diisocyanate. Depending on the OH content of the triglyceride, glassy and elastomeric materials were prepared and characterised fully. Depending on final treatment of the aqueous dispersions, macromolecules bearing negative and positive charges could be obtained readily, thus conferring a wide range of coating applications with good adhesion as a function of the nature of the substrate. These PU are free from volatile organic compounds and are, therefore, 'ecologically sound'.

Larock and co-workers [45] conducted a comprehensive study on an important and hitherto neglected aspect concerning the structure–property relationship associated with residual unsaturations on PU based on vegetable oil epoxy monomers. The authors synthesised a series of hydroxylated epoxidised oils ensuring that the number of OH groups per triglyceride remained constant while varying the extent of their unsaturation (between 0.4 and 3.5 double bonds per triglyceride) using epoxidised substrates (peanut, soybean, linseed, corn and castor oils) and different oxirane-opening reagents. **Scheme 4.12** illustrates this procedure and the fact that different reagents were used alternatively to open the oxirane rings. Polyols were characterised thoroughly by nuclear magnetic resonance spectroscopy, matrix-assisted laser desorption/ionisation-time-of-flight mass spectrometry, and determination of hydroxyl number. Then, they were submitted to water-based polycondensation using isophorone diisocyanate as the cycloaliphatic crosslinker and dimethylol propionic acid as the anionic comonomer. The molar ratio of the reactive groups of the three monomers was kept constant for all polymerisations at 1.0:1.7:0.69.

Characterisation of the ensuing PU after coagulation and drying included static and dynamic mechanical properties, T_g, and thermal stability [45]. Increasing extents of residual unsaturation led to increases in mechanical properties (in MPa), with Young's Modulus increasing from 4.64 to 16.15, toughness from 4.25 to 12.72, and break strength from 2.27 to 7.55, when going from a peanut oil-based to linseed oil-based polyol. Correspondingly, a progressive decrease in percentage strain at break from 343 to 280 was recorded, whereas the T_g increased from −9.2 °C to 13.5 °C. Other factors also played a part, such as the nature of the starting oil and mode of oxirane ring-opening associated with the structure of the appended OH-bearing group. The importance of this investigation resides in having identified specific structural roles in a rigorous fashion, thereby establishing clear-cut criteria for the appropriate choice of a given system as a function of the required application.

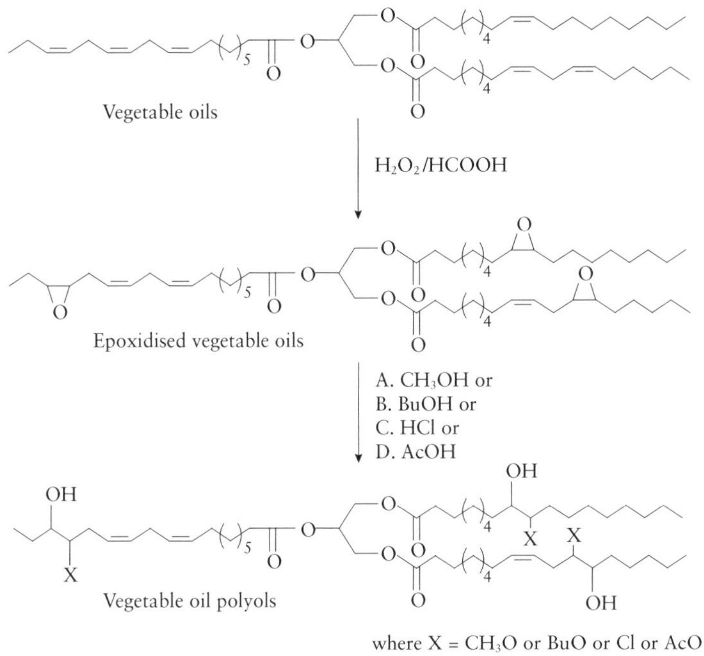

Scheme 4.12 Synthesis of several vegetable oil-based polyols with residual unsaturations [45]

4.3 Polyisocyanates for Polyurethanes

PU are among the most important macromolecular materials because of their versatile properties in numerous technological areas including, thanks to their good biocompatibility and mechanical properties, biomedical applications. It is, therefore, not surprising that, as mentioned above, plant oil-based PU are among the most sought-after materials among polymers derived from those renewable resources.

After having discussed trends in the synthesis and polymerisation of polyols prepared from plant oils, it is now appropriate to examine the state-of-the-art for the synthesis of the corresponding polyisocyanates, i.e., the complementary monomers used to obtain conventional PU. Up to 2005, this topic had been covered only modestly compared with the abundant research on polyols from vegetable oils discussed above, but a new momentum is gathering strength. New studies target synthetic routes that focus on safety concerns and green connotations, which are not found in standard industrial processes that employ phosgene to convert fossil-based amines into isocyanates. These routes are displayed in **Scheme 4.13** [46].

Soybean oil iodo isocyanate was synthesised by reacting the pristine oil with iodoisocyanate (**Scheme 4.13a**) to yield a degree of triglyceride substitution of ≈3 [47]. The same oil was treated with *N*-bromosuccinimide at the allylic positions of each of its fatty-acid chains, followed by reaction with silver isocyanate (AgNCO) (**Scheme 4.13b**) to prepare the desired multi-isocyanate in which ≤70 % of bromine atoms were converted into NCO groups [48].

Diisocyanates derived from fatty acids have been the subject of several studies [49]. Oleic acid (OA) was first converted into a diacid, which in turn was treated in different ways to give 1,7-heptamethylene diisocyanate and 1,16-diisocyanatohexadec-8-ene (**Scheme 4.13c**) [50, 51]. PU prepared with the former diisocyanate displayed physical properties akin to those of the petroleum-derived 1,6-hexamethylene diisocyanate. Counterparts prepared from the latter OA-based diisocyanate exhibited an even higher tensile strength than those derived from petroleum-based 1,7-heptamethylene diisocyanate. The PU prepared with the longer-chain diisocyanate had a lower Young's Modulus and a higher elongation at break because of enhanced chain flexibility. Route **d** in **Scheme 4.13** was applied to dimethyl sebacate [49] and the ensuing diisocyanate used to prepare thermoplastic polyurethane (TPU) elastomers. In all these investigations, use of difunctional monomers and the presence of long aliphatic chains in the ensuing linear polymers resulted in low-T_g materials, some of which were based entirely on renewable resources and some were suitable for biomedical applications.

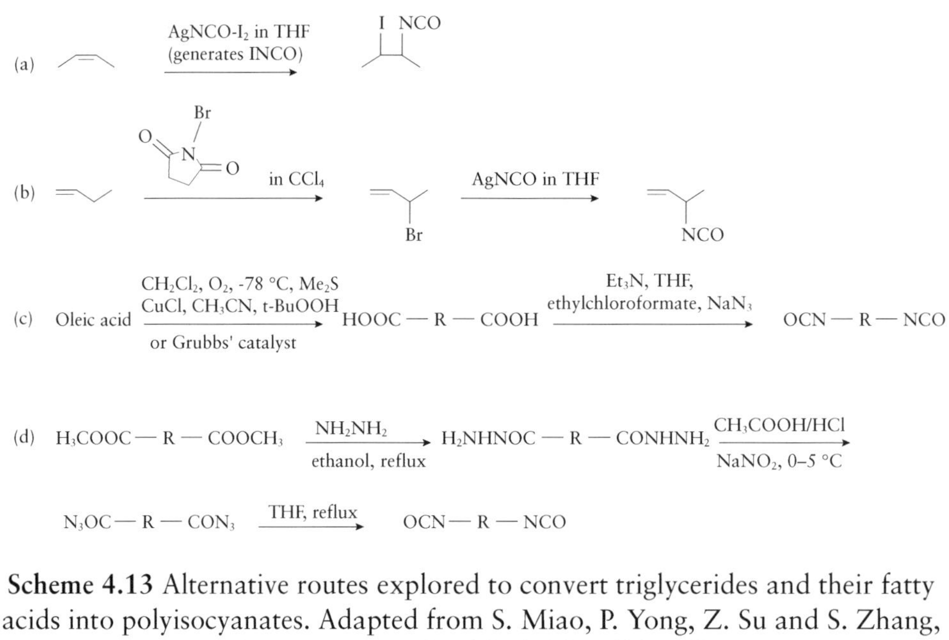

Scheme 4.13 Alternative routes explored to convert triglycerides and their fatty acids into polyisocyanates. Adapted from S. Miao, P. Yong, Z. Su and S. Zhang, *Acta Biomaterialia*, 2014, **10**, 1692 [46]

4.4 Polyether and Polyester Diols for Thermoplastic Polyurethanes

Most of the work discussed above is related to crosslinked PU. However, diisocyanates prepared from the fatty acids of plant oils can also be utilised as comonomers for the synthesis of linear TPU by associating them with diols and preferably from diols derived from renewable resources. TPU display good transparency, tunable stiffness, good wear resistance, and excellent biocompatibility provided their structure incorporates alternations of soft and hard segments in their linear chains. Soft segments are usually low-glass transition oligoethers or oligoesters diols derived from petroleum-based chemistry, but can also be obtained from derivatives of fatty acids. Hard domains are usually crystalline with high melting points, and arise from the use of short diols and/or diamine comonomers.

Only one example of TPU based on a polyether diol from derivatives of vegetable oils has been reported: the structure shown in **Scheme 4.2** [9]. This structure was polymerised with a conventional aromatic diissocynate. The ensuing material had a T_g of 15 °C and good thermal stability, with only 5% weight loss at 305 °C.

The situation concerning polyester polyols is more encouraging. Use of ricinoleic acid as a precursor for these structures is dealt with in **Section 4.9**, which is devoted to the special features associated with this fatty acid and its triglyceride. An oligoester diol with a molecular weight of 1,500 Da arising from the transesterification of methyl 12-hydroxy stearate with 1,6-hexanediol was used as a comonomer in association with an aromatic diisocyanate and1,4-butanediol (1,4-BD) to prepare a segmented PU with T_g values of –40 °C and 100 °C for soft and hard segments, respectively [i.e., a thermoplastic elastomer (TPE)] [52].

In a different vein, polyester diols produced by transesterification were employed in conjunction with an aliphatic diisocyanate and 1,3-propanediol in the synthesis of high-molecular-weight segmented PU [53]. This study involved investigation of the effect of the size of the soft segments on the morphology and physical properties of materials. Transesterification of commercial dimerised fatty acids with aliphatic diols [54, 55] produced a polyester diol with dangling chains that was used for the preparation of segmented PU whose properties could be optimised in terms of specific applications by appropriate 'tuning' of the proportions of hard and soft segments.

Various diester diols (**Scheme 4.14**) were also synthesised by transesterification of methyl oleate and methyl undecenoate with different diols, followed by ring-opening of epoxidised structures [56].

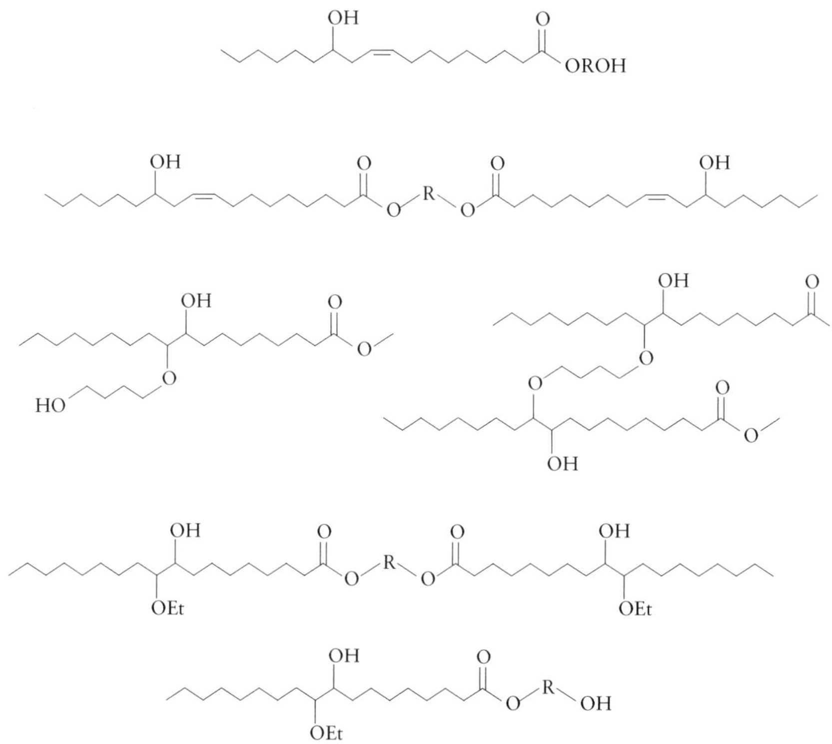

Scheme 4.14 Structures of diester diols from the study by Cramail and co-workers [56]

Their implication in the synthesis of linear PU using an aliphatic diisocyanate produced materials with a T_g below RT, in part because of the flexibility of both monomer units and, additionally, because of the dangling chains in some of the diester diols. Given that these monomers bore one or two secondary OH groups, polymerisations were rather slack and the ensuing molecular weight relatively low.

Sugar-based fatty acid ester diols have also been prepared by transesterification of epoxidised oleates with methyl α-D-glucopyranoside and sucrose, followed by hydrolysis of the oxirane ring [57]. These fully bio-based monomers were polymerised with an aliphatic diisocyanate to produce PU whose structure could be oriented toward a linear architecture (when the sugar OH groups were not involved) or a network (if at least some of them participated in the polycondensation) by changing the solvent medium.

Other interesting structures belonging to this family of monomers were also synthesised by thiol-ene chemistry and by using ricinoleic acid (see **Chapter 6**).

4.5 Diols and Diacids for Linear Polyesters

Aliphatic thermoplastic polyesters represent a class of materials that is attracting a considerable amount of attention because they are i) biodegradable and biocompatible and ii) increasingly accessible from the exploitation of diols and dicarboxylic acids derived from renewable resources. If long methylene chains are present in the monomers, the ensuing products resemble polyethylene (PE) in structure and, hence, in most properties, have the added advantage of biodegradability.

Long-chain α,ω-hydroxy acids, as well as long-chain diols and diacids isolated from the hydrolysis of suberin, are good examples of monomers from renewable resources that have been used to synthesise polyesters [58, 59]. Likewise, triglycerides from vegetable oils, with their aliphatic structure and ability to be spliced into three linear strands are, therefore, extremely suitable substrates for the synthesis of both complementary difunctional monomers used for the preparation of long-chain aliphatic polyesters. Among the several mechanistic approaches applied for this purpose to convert fatty acids and esters from natural triglycerides, the very important thiol-ene reaction, metathesis routes, and use of ricinoleic acid are discussed in more specific contexts in **Chapters 5** and **6**.

Alkylcarbonylation is an interesting approach for the preparation of diesters from derivatives of fatty acids. Methyl oleate and methyl erucate were reacted with carbon monoxide at high pressure and temperature in methanol and ethanol using palladium-based catalysts to prepare the respective C19 and C23 diesters [60–62]. Their reduction and hydrolysis gave the corresponding diols and diacids, which were employed for the synthesis of 19,19 and 23,23 long-methylene chain polyesters. Their melting temperature (T_m) values were 103 and 99 °C, respectively. In a similar vein [63], olive, rapeseed and sunflower oils were submitted to methoxycarbonylation using an efficient palladium-based catalyst to synthesise the corresponding diesters in high yields, which were then used to prepare the corresponding diacids and diols (**Scheme 4.15**).

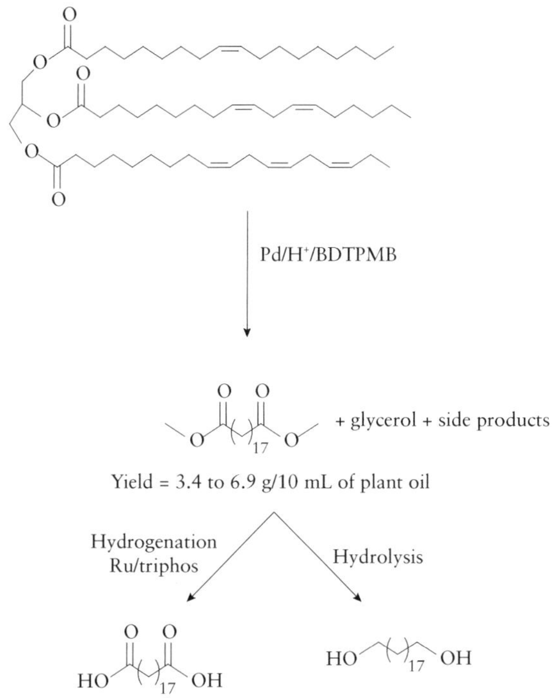

Scheme 4.15 Transformation of plant oils into their corresponding diesters, diols and diacids. BDTPMB: Bis(ditertiarybutylphosphinomethyl)benzene [63]

12-Hydroxydodecanoic acid, prepared by the oxidative cleavage of vernolia oil was used directly as an AB monomer to synthesise a high-molecular-weight polyester with a T_m of 88 °C [64]. Fatty acid-derived ω-pentadecalactone (PDL) was polymerised efficiently by an enzyme-catalysed ring-opening mechanism to give a polyester [poly(ω-pentadecalactone) (PPDL)] with a T_g of –27 °C and a T_m of 97 °C [65–67]. Similarities between this polymer and PE were examined by comparing its properties as a function of molecular weight to those of high-density polyethylene (HDPE) [68]. Similar values for Young's Modulus and stress at yield were encountered, but the elongation at break was higher for the polyester and more akin to that of linear low-density polyethylene (LLDPE), probably due to the C–O bonds in its chains. Fibres were also melt-spun using high-molecular-weight samples of this polyester and, after elongation, their mechanical properties were enhanced thanks to a high degree of crystal orientation [65–67].

Chemical catalysis, which had failed in previous studies, was applied successfully recently to the ring-opening polymerisation (ROP) of PDL using cheap metal-based initiators [69, 70] and organic counterparts [71]. **Scheme 4.16** summarises the alternative systems employed to polymerise this cyclic monomer.

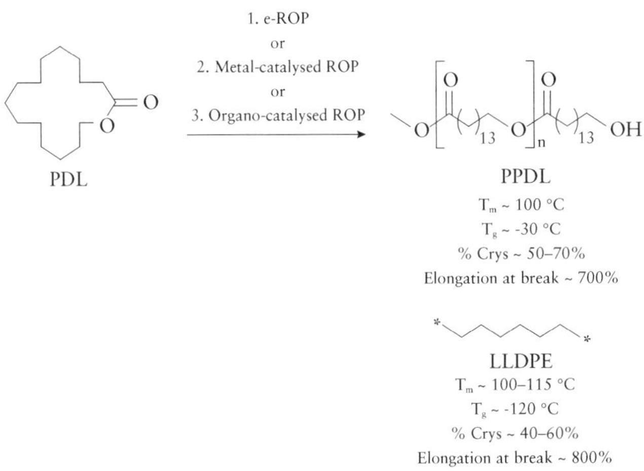

Scheme 4.16 Three routes leading to the ROP of PDL. e-ROP: Enzymatic ring-opening polymerisation [71]

C12 and C16 fatty acids were used to prepare long-chain hydroxy acids by a biosynthetic approach, and the methyl ester of ω-hydroxytetradecanoic acid was selected as a test AB monomer [72]. Its polymerisation induced by Ti(OiPr) was studied thoroughly, together with the properties of the ensuing polyesters of varying molecular weight. Again, strong similarities with HDPE were found.

The search for polyesters derived from plant oils that mimick the properties of PE has progressed considerably, with more success for simulating the behaviour of the amorphous low-density material because generating crystalline morphologies similar to those of the high-density counterpart is difficult. The qualitative difference, however, resides in the fact that one expects these long aliphatic chain polyesters to be biodegradable, with obvious positive ecological and biomedical implications.

A different strategy providing access to linear polyesters derived from fatty acids in which their aliphatic sequences dangle from macromolecular chains (i.e., a type of structure very different to that of linear PE) has been described recently [73]. Various saturated fatty acid methyl esters were malonated to the corresponding methyl diesters, which were polymerised by transesterification with 1,6-hexanediol (**Scheme 4.17**).

Polyesters with side chains bearing 6 or 8 methylene groups were amorphous with T_g values below –60 °C. The other polymers displayed some crystallinity and their T_m increased with increasing side-chain size <27 °C, with crystallinity being attributed to the intermolecular organisation of these dangling aliphatic strands.

Presence of reactive moieties along polyester chains represents an interesting challenge which could yield original properties and the possibility of specific chemical modifications of these macromolecules. Thus, for example, linear (or at least not crosslinked) polyhydroxyesters can be prepared by the reaction of a diacid with a triol providing that the criteria *viz.* correct stoichiometry and use of triols with tunable reactivity with respect to their different OH groups are met.

Within the present context, an oleic-inspired, dicarboxylic acid, 1,18-*cis*-9-octanedecenedioic acid, was synthesised by enzymatic means. Thereafter, it was used as a monomer in conjunction with glycerol to prepare a thermoplastic polyester incorporating OH functions by enzyme catalysis promoted by lipase B from *Candida antarctica* [74]. Given the pronounced difference in reactivity between the primary and secondary OH groups of glycerol, coupled with a starting 1:1 monomer molar ratio, the ensuing polymer architecture was essentially linear with a modest contribution from branching. Hence, the polyester bore about one secondary OH group per monomer unit.

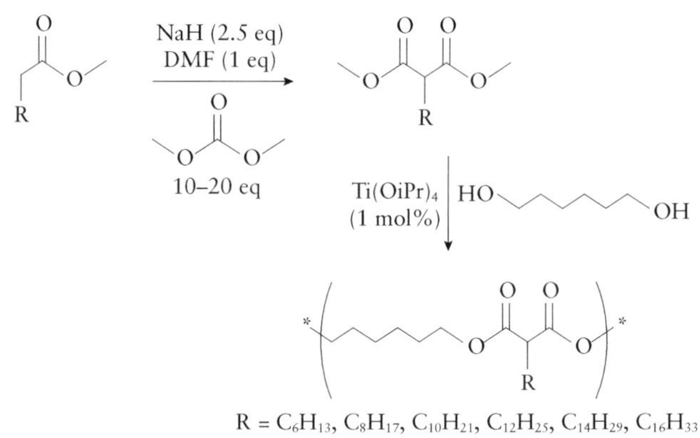

Scheme 4.17 Synthesis of linear polyesters with dangling aliphatic chains using different fatty acids. DMF: Dimethylformamide [73]

Very different macromolecular structures were obtained in an investigation in which divinyl sebacate, glycerol and unsaturated or epoxidised fatty acids were made to polymerise by transesterification and direct esterification under the action of *Candida antarctica* lipase (CAL) [75, 76]. When unsaturated acids were used, their C=C moieties in the ensuing polymers were later epoxidised to form oxirane rings. These polyesters were characterised by the fact that they bore the fatty acid-based chains as side strands formed by the reaction of their carboxylic functions with the secondary OH moieties of glycerol with respect to the linear chains arising from the reaction of the diester with the primary hydroxyl groups of glycerol. Subsequently, films of these materials were crosslinked to prepare biodegradable artefacts. Epoxy moieties along the linear chain of a polyester based on vegetable oils was achieved through use of a diacid bearing one oxirane group that was polymerised with various aliphatic diols (**Scheme 4.18**) [77]. These polyesters, prepared by enzymatic catalysis, displayed relatively high molecular weight of tens-of-thousands of daltons.

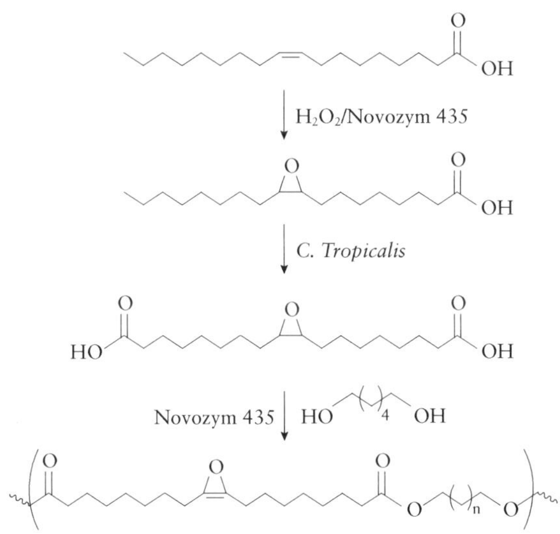

Scheme 4.18 Linear aliphatic polyesters incorporating epoxide functions [77]

In a different vein, 10,11-epoxyundecanoic acid was exploited as an AB monomer, which was polymerised by the oxirane ring-opening chain process induced by the carboxylic group sitting at its other end and catalysed by ammonium or phosphonium salts [78], as illustrated in **Scheme 4.19**. The ensuing linear hydroxylated polyester exhibited mechanical properties similar to those of linear LDPE, but with better adhesion capacity thanks to the OH groups and a strong barrier against oxygen.

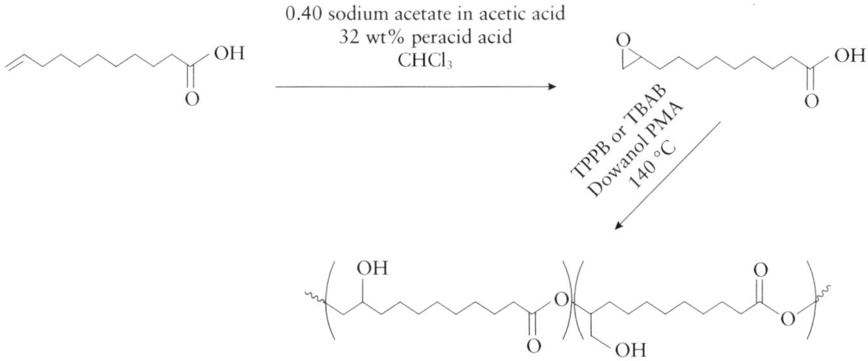

Scheme 4.19 Homopolymerisation of an epoxide-carboxylic monomer. PMA: Propylene glycol monomethyl ether acetate; TBABr: tetrabutylammonium bromide; and TPPB: tetraphenylphosphonium bromide [78]

Self-polymerisation of epoxidised OA through a similar mechanism, but this time achieved simply by heating the monomer, produced a polyhydroxyester with aliphatic side chains bearing seven carbon atoms [79]. Careful structural characterisation of this material confirmed the chain-polymerisation mechanism based on COOH-induced opening of the oxirane ring to give ester moieties within the chain and OH side groups. This polyester polyol was utilised as a macromonomer for preparation of crosslinked PU in conjunction with diisocyanates. When diglycidyl sebacate was polymerised by the same ring-opening mechanism (catalysed by sebacic acid), a branched polyhydroxyl ester was obtained whose OH groups were subsequently exploited to append bioactive groups by esterification [80] with a high yield of functionalisation.

Zhang and co-workers developed unique crosslinkable unsaturated polyesters by a one-pot lipase-catalysed oxidation followed by copolymerisation of crude linoleic acid, glycerol and 1,18-*cis*-9-octanedecenedioic acid [81]. In a similar approach, Roumanet and co-workers prepared aliphatic unsaturated polyesters derived from 1,18-(Z)-octadec-9-enedioic acid (a derivative of fatty acids from sunflower oil) and aliphatic diols of different molecular weight [82]. The double bonds of the ensuing polyester backbones were then epoxidised and photo-crosslinked, yielding transparent and homogeneous crosslinked films with high hydrophobicity.

4.6 Monomers for Linear Polyamides and Polycarbonates

The importance of thermoplastic polyamides (PA) in the realm of macromolecular materials is reflected in their wide range of applications in numerous technological domains. Their excellent mechanical properties arise from intermolecular N-H···O=C hydrogen bonding, which also facilitates polymer crystallisation. The standard way of representing their structure in the case of straightforward aliphatic sequences arising from the diacid and diamine precursors is to give the number of carbon atoms of each preceded by the PA abbreviation. This can be shown by the most widespread material: PA-6,6 (Nylon). Curiously, the number of PA produced by industrial means is not large and, moreover, the linear structures dominate as market commodities. The interest in PA based on fatty acids dates to the 1960s. However, recent work has tackled a wider variety of approaches, such as use of commercial acid dimers in conjunction with various diamines [83, 84]. Such contributions to this field, in which the diamine or diacid or, better still, both monomers, were based on vegetable oils, are outlined here.

Polycondensation of adipic acid and 1,13-tridecane diamine afforded PA-13,6 with a T_g of 60 °C and a T_m of 206 °C, *viz.* viable transitions for certain applications [85, 86]. Preparation of pure long-chain diacids from methyl oleate and erucate has been mentioned above [60–62] and in this comprehensive investigation the authors also prepared the corresponding C19 and C23 diamines by the four-step reaction sequence shown in **Scheme 4.20.**

Their association with the corresponding diacids produced PA with a molecular weight of ≈10 kDa [61]. PA-23,19 and 23,23 displayed a similar T_m ≈155 °C. An interesting diamine, nonadecane-1,19-diamine, derived from sunflower oil by a set of reactions going through the corresponding diol, was synthesised, characterised, but not used as a PA precursor [87]. A different strategy was adopted to prepare thermoplastic PA from derivatives of α-arylated fatty acids [88]. The reaction of 1,4-dibromobenzene and fatty acid *t*-butyl esters was followed by preparation of the corresponding carboxylic acids by hydrolysis of ester functions. These dicarboxylic acids were then polymerised with hexamethylene diamine to yield PA with relatively low molecular weight (probably because of steric hindrance associated with the structure of the diacids). In another attempt to arrive at viable PA [89], bromination of methyl oleate and methyl erucate was followed by the reaction with sodium azide (NaN_3) and subsequent reduction of the ensuing product with hydrogen to give AB-type monomers bearing ester and amino groups. Their polymerisation gave PA with a reasonable molecular weight of 6–8 kDa and a T_m of 40 and 89 °C for the methyl oleate and methyl erucate-based materials, respectively.

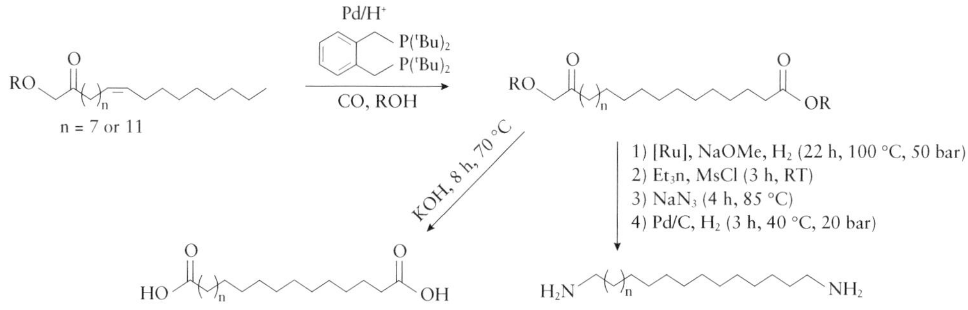

Scheme 4.20 Synthesis of diacids and diamines from vegetable fatty acid methyl esters [60–62]

Other publications dealing with the context of PA derived from vegetable oils are discussed in **Chapters 5** and **6** dealing, respectively, with metathesis and thiol-ene reactions.

Polycarbonates (PC) are among the most prized engineering materials because of their stability to all types of degradation (except biodegradability for aliphatic homologues), transparency, mechanical properties and thermal resistance, though their uses are confined to high-tech or biomedical applications because of their relatively high price. Syntheses and characterisation of PC derived from vegetable oils are not numerous, and most of the monomer preparations are based on reaction mechanisms that are discussed elsewhere in this book.

A recent contribution by more conventional chemistry describes the synthesis of 1,10-dodecanediol by the reduction of sebacic acid, followed by its conversion into the corresponding α,ω-bis-carbonate by reacting it with methylchloroformate or dimethyl carbonate [90]. Carbonate interchange reactions were then applied to polymerise this monomer in the melt using a choice of catalysts, and the ensuing PC had a large domain of molecular weight (8–50 kDa). An alternative strategy called upon direct polymerisation using the original diol and dimethyl carbonate. These linear polymers had relatively low T_g values owing to long aliphatic sequences in their chains (e.g., 58 °C for the semi-crystalline member incorporating 10 methylene strands).

4.7 Vinyl, Acrylic and Other Monomers for Linear Chain-growth Polymerisation

The synthesis of monomers derived from fatty acids incorporating end-functions susceptible to classical polyaddition reactions has been revived recently and yielded interesting results. Vinyl oleate (VO) and vinyl linoleate (VL) were synthesised by a transvinylation reaction of the corresponding fatty acids with an excess of vinyl acetate (VAc) in bulk using a safe iridium-based catalyst instead of previously used mercury-based counterparts (which are not acceptable today because of safety concerns) [91]. **Scheme 4.21** summarises this synthesis in the case of OA.

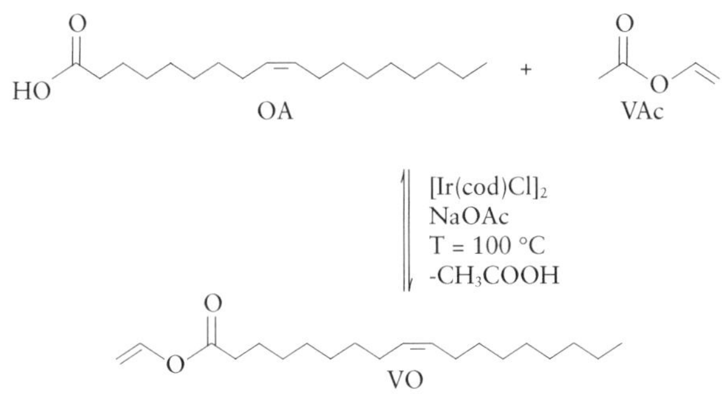

Scheme 4.21 Synthesis of VO by transvinylation [91]

Free-radical polymerisation of these monomers as well as their copolymerisations with VAc with variable monomer feed ratios (but always with an excess of VAc) were conducted under standard conditions using benzoyl peroxide (BP) as initiator, and produced materials with the expected structures [91]. The reactivity of the vinyl fatty-acid compounds was found to be essentially the same as that of VAc. Their homopolymerisations gave oligomeric products because of the steric hindrance associated with the propagation step, thus favouring transfer reactions. The weight average molecular weight of the copolymers reached 8 kDa with a polydispersity index (PDI) ≈2. Close inspection of the structure of these macromolecules clearly showed that the unsaturations in the fatty-acid chains had not been involved in the processes and that the random copolymers had the structure shown in **Scheme 4.22** for copolymerisation of VL with VAc.

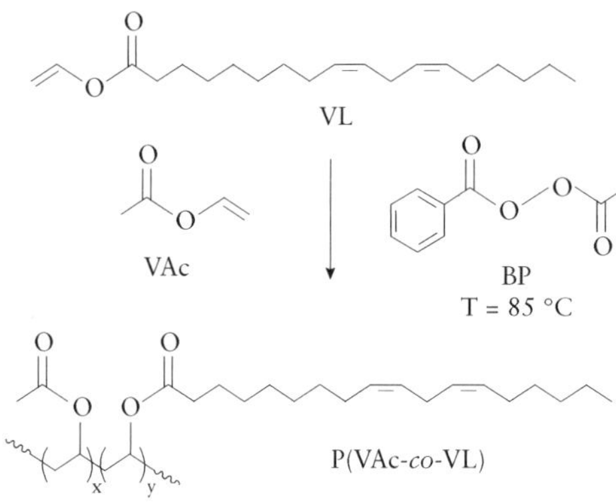

Scheme 4.22 Random free-radical copolymerisation of VL with VAc

The homopolymers displayed low T_g values: –15 for poly-VO and –50 °C for poly-VL. This difference was attributed to the higher mobility of the dangling chains of the latter because of their higher degree of unsaturation. A 50/50 copolymer prepared with these two fatty-acid monomers gave a T_g of –35 °C (i.e., halfway between those of the corresponding homopolymers), thereby confirming the random character of its structure [91]. The copolymers exhibited T_g values ranging from –5 to 16 °C as a function of their increasing composition of VAc units.

The purpose of this study stemmed from considerations related to the atmospheric drying of inks, paints and coatings based on vegetable oils. Structural and tactile evolution of films of the copolymers was followed by Fourier-Transform infrared spectroscopy and tack measurements. They showed the progressive drying features associated with the classical pattern of progressive branching and final crosslinking (as shown in **Scheme 2.1** in **Chapter 2**) induced by the free-radical process of oxygen attack onto the C=C unsaturations in the dangling chains of the fatty acid monomer units. This process was accelerated strongly by addition of conventional catalysts based on soluble acetates of calcium, cobalt and zirconium [91]. It appears, therefore, that these materials, derived in part from renewable resources, could find useful applications as paints and coatings without volatile components.

Many publications dealing with the free-radical homo/copolymerisation of saturated fatty acid acrylates and methacrylates appeared between 2001 and 2011, particularly using living systems such as atom transfer radical polymerisation (ATRP) [91–112]. Monomers were prepared, for example, by the reaction of acrylic and methacrylic acid chlorides with fatty alcohols of different chain length, as shown in **Scheme 4.23** in the case of methacrylates (which also includes their ATRP conditions). A very broad spectrum of conditions have been scanned in these numerous studies for homopolymerisations of the different monomers as well as random and block copolymerisations, mostly with conventional acrylates and methacrylates, but also with styrene and caprolactone, among others.

These investigations revolve around homopolymers with different sizes of the dangling chains in terms of thermal properties, hydrophobic character and possible applications. Thus, for example, the tendency of the side chains to associate in an organised fashion produces partial crystallisation of the polymers which, together with their hydrophobicity, can enable their use as nucleating agents, rheology modifiers, dispersion stabilisers, and smart gels. Also, using hard-soft block copolymers with methyl methacrylate, thermoplastic or elastomeric properties could be modulated as a function of their composition.

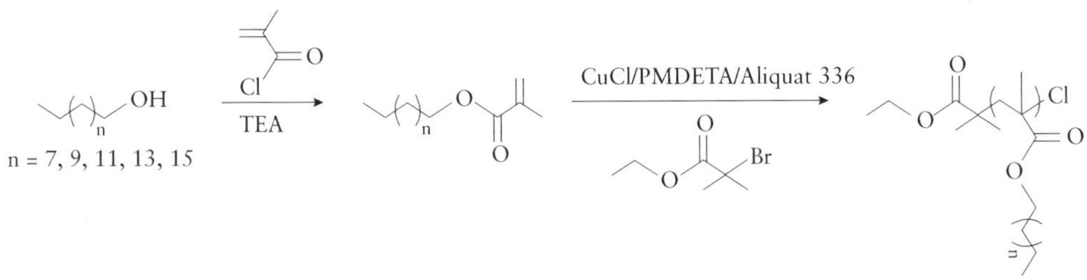

Scheme 4.23 Synthesis of saturated fatty acid methacrylates and their atom transfer radical polymerisation. PMDETA: *N,N,N',N',N''*-Pentamethyldiethylenetriamine and TEA: triethylamine [110]

The interested reader can find a more detailed discussion about these investigations by consulting the section devoted to them in a very thorough review by the research team of Cramail [113].

A different strategy was proposed recently by Rangasami and Mannari [114] who i) repeated incompletely the study by Pelletier and co-workers [25, 26] on the acrylation of epoxidised oils but ii) also studied their reaction with hydroxyethyl acrylate in the presence of a strong acid to prepare the corresponding polyacrylate ethers for coating applications, as described previously [25, 26].

All of the studies described above dealt only with *saturated* fatty-acid acrylic monomers except for brief reports on the preparation of oleyl methacrylate [115, 116].

In a more comprehensive ongoing investigation [117], fatty acid-based monomers with acrylate functionality were prepared under *mild and green conditions* by direct esterification of *unsaturated* fatty acids, namely 10-undecenoic (C11:1), oleic (C18:1) and linoleic (C18:2) acids, with 2-hydroxyethyl methacrylate (HEMA) in the presence of *N*,*N*'-dicyclohexylcarbodiimide and 4-dimethylaminopyridine. **Scheme 4.24** shows the structure of these novel monomers, whose additional originality is that they were synthesised *directly* from the corresponding fatty acid.

The monomers, obtained in near-quantitative yields, were submitted to free-radical homopolymerisation and copolymerisation with HEMA using 2-hydroxy-2-methyl-1-phenyl-propan-1-one as the photo-initiator. The interest of these materials (as opposed to the saturated counterparts reviewed above) is reactive C=C moieties within and at the end of their dangling chains. Their properties and chemical modifications are being investigated.

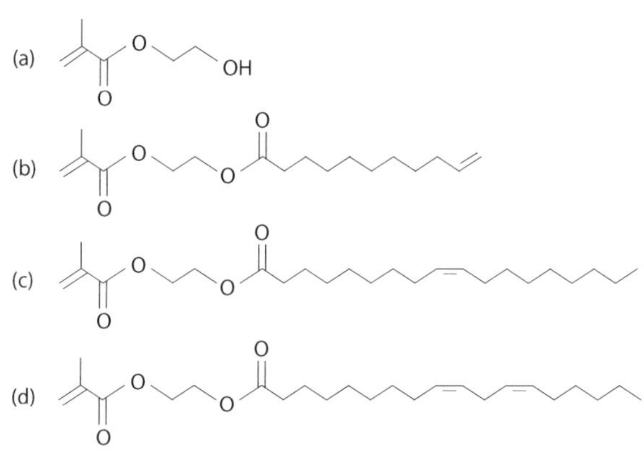

Scheme 4.24 The three unsaturated fatty-acid methacrylates (b, c, d) and the alcohol used for their synthesis (a) [117]

The other family of fatty acid-based monomers for chain-growth polymerisations that has attracted considerable attention is oxazolines. These heterocyclic structures are polymerised readily by cationic ring-opening processes [118]. The ensuing polyoxazolines have found numerous applications [119] as in biomedicines [120, 121] and as non-ionic surfactants or emulsifiers, among others.

Fatty-acid chains with oxazoline end groups have been synthesised by the two mechanistic routes shown in **Scheme 4.25** and their polymerisations and copolymerisations reviewed [122]. Block copolymers of conventional alkyl oxazolines with comonomers like those shown in **Scheme 4.25** have been used to exploit their amphiphilic properties. When the fatty-acid chains incorporated C=C unsaturations, further chemical modifications were introduced to obtain materials for more specific applications [123] (e.g., incorporation of hydroxyl groups for the synthesis of PU networks). Overall, however, the most important use of these polymers and copolymers is non-ionic surfactants with specific abilities to form micelles with original morphologies, which can function in many interface-related applications, such as micellar catalysis.

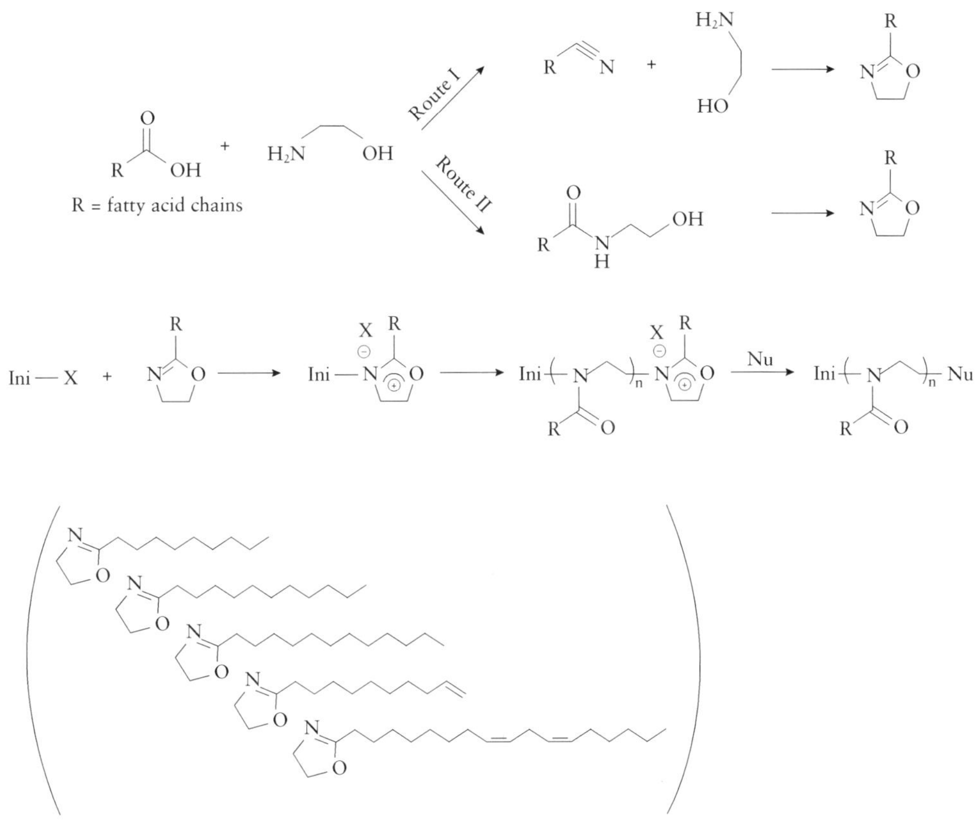

Scheme 4.25 Two ways of preparing fatty acid-bearing oxazolines, and their cationic polymerisation [122]

4.8 Monomers for Other, Less Common Linear Polymers

There have been few recent studies on the synthesis of polyethers and polythioethers derived from vegetable oils, and all arise from castor oil-based monomers. They are, therefore, discussed in the next section devoted to this very special renewable resource. The same applies to polyanhydrides.

4.9 Special Cases of Castor Oil and Ricinoleic Acid

The fact that castor oil is a well-established industrial commodity because of the importance of some of its derivatives and macromolecular materials explains why so many studies on polymers derived from vegetable oils have focused on the use of ricinoleic acid, 10-undecenoic acid, and several other molecules based on their structures. **Chapter 3** considered the context of *direct* polymerisation of this oil and its main fatty acid. This section is devoted to reviewing the most salient recent aspects of the vast realm of monomers and polymers based on their second-generation derivatives. Its treatment begins after the exhaustive review that Mutlu and Meier published in 2010 [124]. The impressive number of contributions since 2010 testifies to the relevance that this area has acquired. Some general notions are, however, needed to 'set the stage'.

The most relevant industrial success of castor oil as a precursor to macromolecular materials is the synthesis of PA 11. **Scheme 4.26** illustrates the chemical steps leading from the oil to this high-tech PA. In this process, some of the intermediates represent in themselves very attractive molecules as purveyors of macromolecular materials (notably 10-undecenoic acid). This phenomenon explains blooming of the involvement of research teams in exploiting that field, quite apart from studies in which the pristine oil and its major fatty acids represent starting commodities. The layout of the section is, therefore, set to deal first with the latter aspect before tackling the most representative publications dealing with the former.

Transesterification of castor oil with triethanolamine or glycerol was carried out to increase its hydroxyl functionality [125]. The ensuing polyols were used to prepare rigid vegetable oil foams with an apparent density of 50 kg/m^3 and a compression strength ≈200 kPa.

Reactivity of the ricinoleic acid oligomer discussed in **Chapter 3** was enhanced by esterifying its terminal secondary hydroxyl group with 10-undecenoic acid (i.e., a second-generation member of the same family) using an enzymatic approach (**Scheme 4.27**) [126]. The terminal vinyl group was deemed much more exploitable than the original OH and COOH moieties for further chemical modifications.

Inclusion of these structures into polyglycerols was met with a moderate outcome because of poor yields.

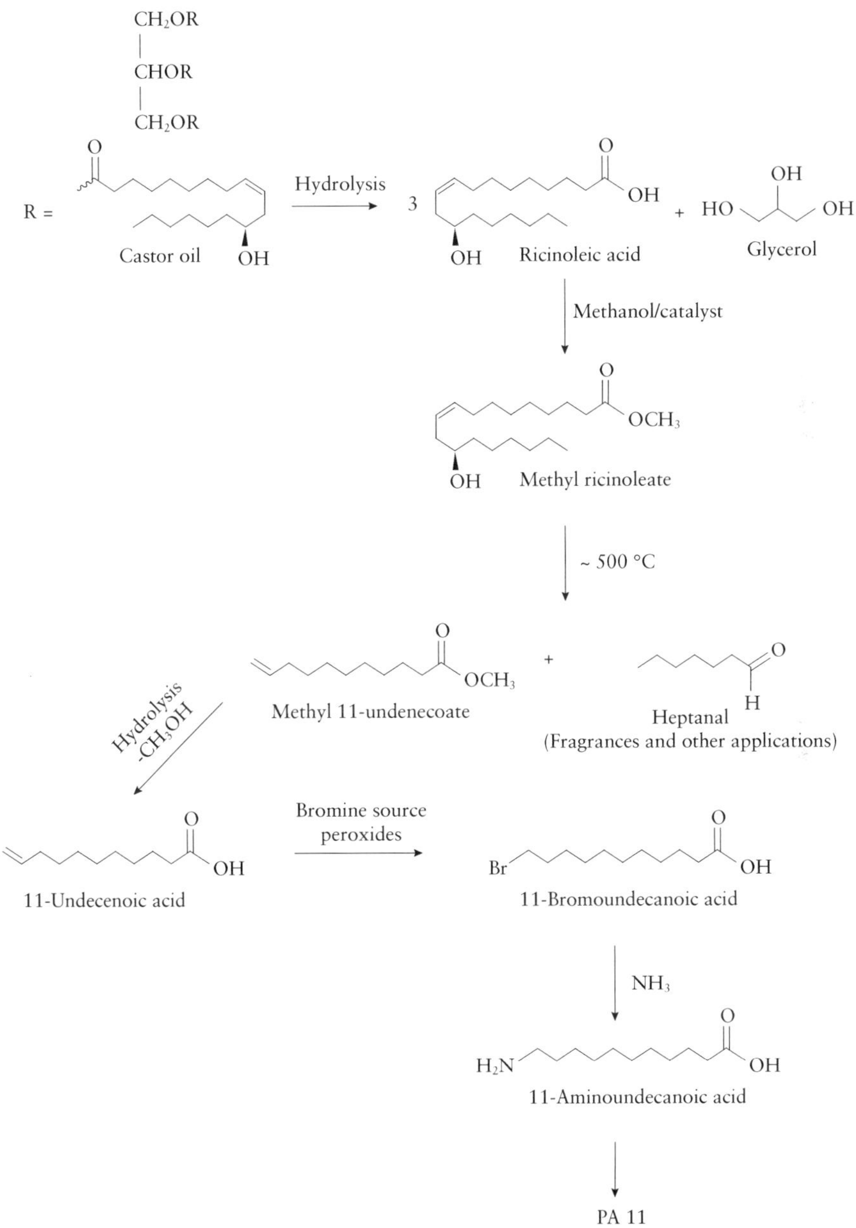

Scheme 4.26 'Chemical flowchart' from castor ricinoleic triglyceride to polyamide (PA) 11

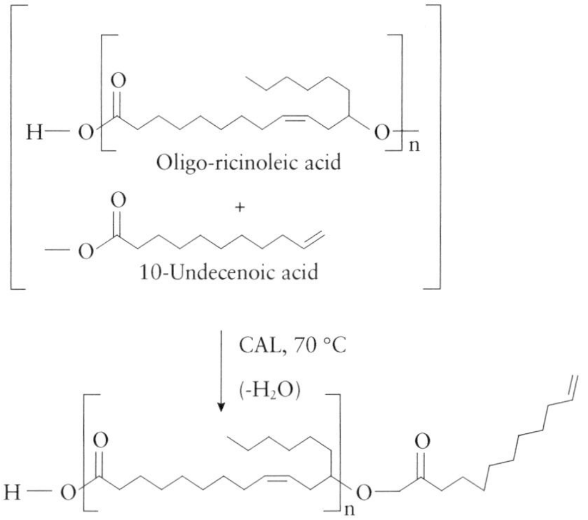

Scheme 4.27 Introduction of a vinyl end-group onto polyricinoleic acid by immobilised *Candida antarctica* lipase [126]

Petrovic and co-workers [127] used a polyricinoleic acid diol of 2,580 Da as a soft segment to prepare a series of PU TPE based on an aromatic diisocyanate and 1,4-BD, with soft-segment sizes ranging from 40 to 70% (**Scheme 4.28**). These materials displayed a microphase separation and two T_g at around –50 and 100 °C. T_m values in the range 170–220 °C were observed depending on the soft-segment content. Their mechanical and thermal properties suggested that they could be used in conventional TPE applications.

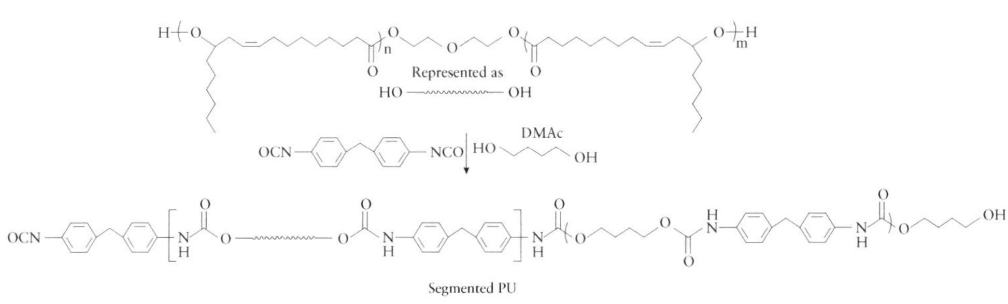

Scheme 4.28 Preparation of segmented polyurethanes from polyricinoleic acid. DMAc: dimethylacetamide [127]

A very recent contribution to materials based on ricinoleic acid [128] revived two main aspects. The first was a search for optimum conditions related to its homopolymerisation and copolymerisation with butylene succinate. This was attained by using titanium tetrabutoxide at >200 °C to yield products with high molecular weight. Secondly, these materials displayed high biocidal activity, which was enhanced further by appending imidazolium moieties to their macromolecules.

Turning to undecenoic structures, the acid, given its availability as an industrial commodity as a product of the pyrolysis of castor oil, coupled with its potentially reactive structure, is a very attractive molecule for approaching new ways of preparing polymers derived from vegetable oils, particularly with research teams working with the new strategies discussed in **Chapters 5–7** (where much of the studies making use of it are reviewed).

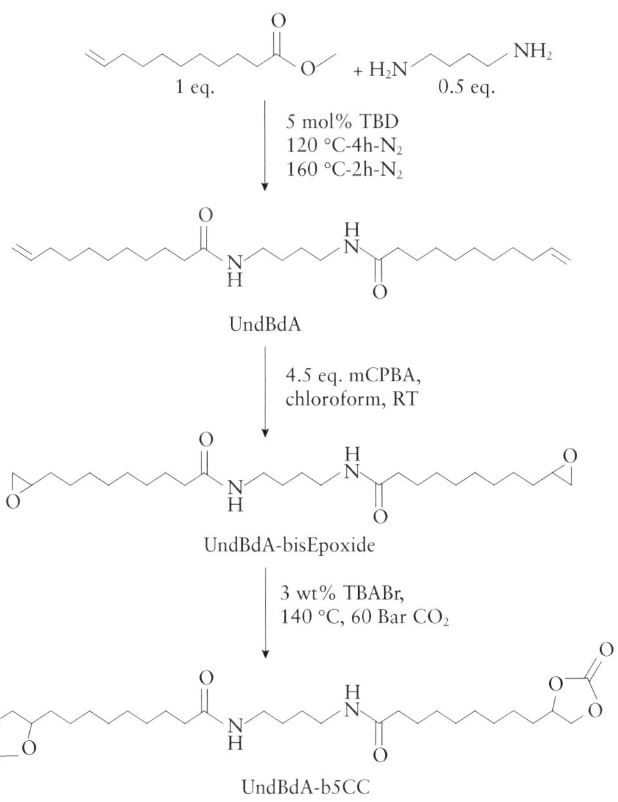

Scheme 4.29 Synthesis of bis-cyclic carbonate macromonomers based on methyl 10-undecenoate. TBD: 1,5,7-Triazabicyclo[4.4.0]dec-5-ene and UndBdA: undecenylbisdiamide [128]

An example of its exploitation by more conventional mechanisms relates to the synthesis of bis-cyclic carbonates [129]. Methyl 10-undecenoate was reacted with various diols and diamines to open the pathway to the synthesis of the corresponding bis-carbonates macromonomers bearing an ester or an amide core, as shown in **Scheme 4.29** for the case of butane-1,4-diamine. These reactive intermediates were then used to prepare polyhydroxyurethanes (PHU) using a wide selection of diamines following the coupling reactions shown in **Scheme 4.30**.

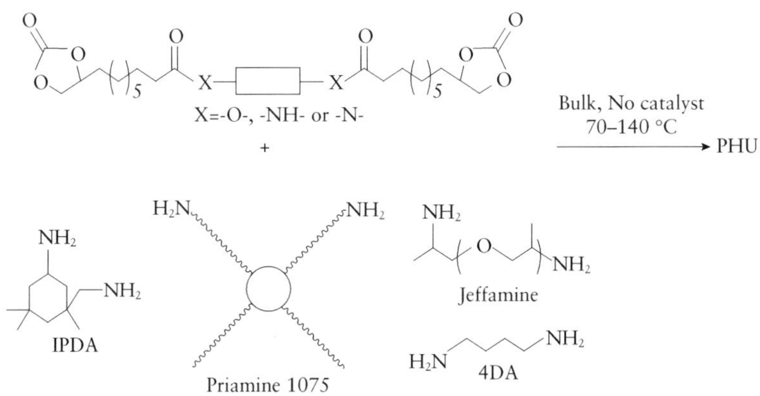

Scheme 4.30 Preparation of various polyhydroxyurethanes. IPDA: Isophorone diamine [128]

This green-chemistry strategy provided the possibility of arriving at urethane moieties without use of standard unsafe methods based on phosgene. Polymers had molecular weights ranging between 11 and 31 kDa, T_g values of –29 to 55 °C and a T_m of ≈110 °C, for those whose structure was sufficiently regular to promote some crystallisation.

The relative paucity of studies discussed in this section does not contradict the emphasis on the topic because many more investigations on castor-oil derivatives appear in **Chapters** 5 and 6, where they make up the bulk of the approaches based on metathesis and thiol-ene processes for new monomer and polymer synthesis.

4.10 Special Case of Glycerol

Worldwide activities related to biodiesel production have placed onto the market enormous quantities of glycerol as a byproduct of the transesterification of triglycerides by methanol or ethanol. In this context, two figures are staggering: the steadily growing worldwide production of glycerol (≈1.2 Mt in 2010) and a steadily decreasing price (≈$0.1/kg). This very active scenario has promoted a host of research activities aimed at looking for novel uses for glycerol, and studying its possible transformations into various compounds, mostly well-known chemical commodities usually derived from fossil resources (including monomers) [130–132]. Before dealing with the latter structures, a succinct discussion of the use of glycerol as a source of macromolecular materials is important.

Glycerol can be polymerised by two basic step-growth mechanisms. These mechanisms lead, respectively, to linear and highly branched oligomers with degrees of polymerisation ranging typically from 2 to 6, depending primarily on the nature of the catalysts, which can be acidic or basic, homogeneous or heterogeneous, as discussed in recent reviews [133]. If etherification takes place only between primary OH groups, as in **Scheme 4.31**, the ensuing oligomers bear a linear structure, whereas participation of the secondary moiety gives rise to highly branched structures (**Scheme 4.32**). Applications of these oligomers include cosmetics, food additives and lubricants.

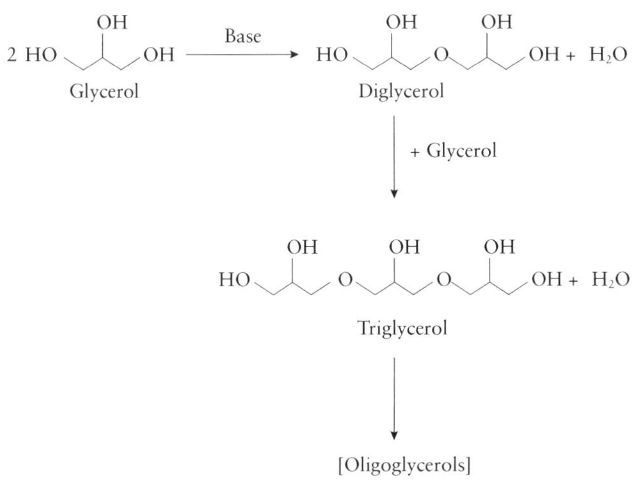

Scheme 4.31 Linear growth of oligoglycerols catalysed by a nucleophilic agent

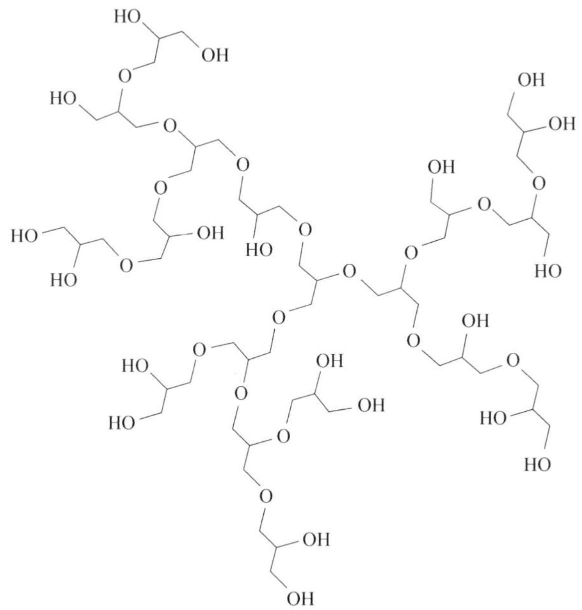

Scheme 4.32 Model structure of a hyperbranched polyglycerol

Synthesis of crosslinked hydrogels using epichlorohydrin (itself derived from glycerol) or PE glycol diglycidyl ether as a co-reactant has been studied [134]. Search for systems leading to high-molecular-weight products is an important current issue [135]. If a single multifunctional supported CeBiPt/C catalyst is used under basic [136] or acidic [137], conditions, glycerol is converted directly into polyketomalonate in an elegant one-pot oxidative polymerisation leading to a high-molecular weight polycarboxylate, which is an excellent building block for household detergents.

The chemistry of glycerol has recently undergone a radical upgrade [130–132, 138, 139]. Among the multiplicity of derivatives investigated in this research effort, some monomers stand out, all well-known structures derived from fossil resources.

Further chemical modifications of glycerol typically involve selective esterification with fatty acids to produce glycerol and polyglycerol esters that can be used as food emulsifiers (e.g., glycerol monooleate) [140] or non-ionic surfactants [141] that can be used as detergents [142]. Production of other chemicals by different reaction pathways has become an additional form of its exploitation [130–132, 138, 139], as illustrated in **Scheme 4.33**, in which numerous conventional monomers derived from petrochemistry also appear.

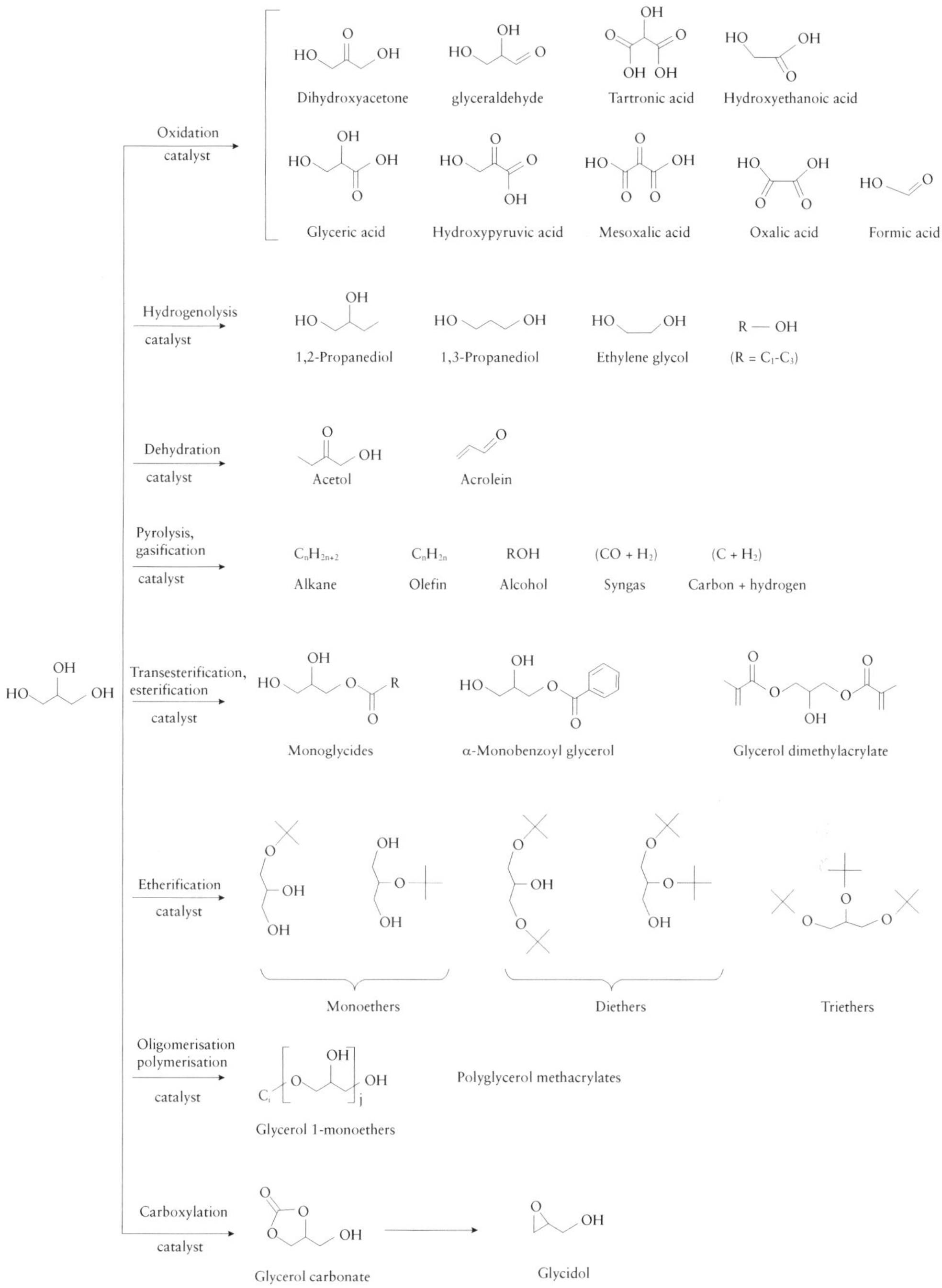

Scheme 4.33 Processes of glycerol conversion into value-added chemicals

Oxidation of glycerol leads to a complex reaction pattern in which many products, such as dihydroxyacetone, glyceric acid, hydroxypyruvic acid, mesooxalic acid and tartronic acid, are formed [138, 139], all of which have great potential as intermediates for the synthesis of polymers [143–148]. Hydroxyesters or hydroxyacids from glycerol have been polymerised by transesterification to prepare biodegradable hyperbranched PC [149] and polyesters [150, 151]. Selective hydrogenolysis of glycerol is a metal-catalysed reaction that can produce 1,2-propanediol, 1,3-propanediol and ethylene glycol (i.e., well-known polycondensation monomers).

Glycerol dehydration is an interesting pathway to acrolein (**Scheme 4.34**), a sustainable and cost-effective alternative route for the commercial catalytic petrochemical process. Acrolein is an important chemical intermediate for the production of acrylic esters, superabsorbent polymers and detergents [152, 153]. Other acrylic monomers based on glycerol have also attracted much attention, as discussed thoroughly in a recent review [154].

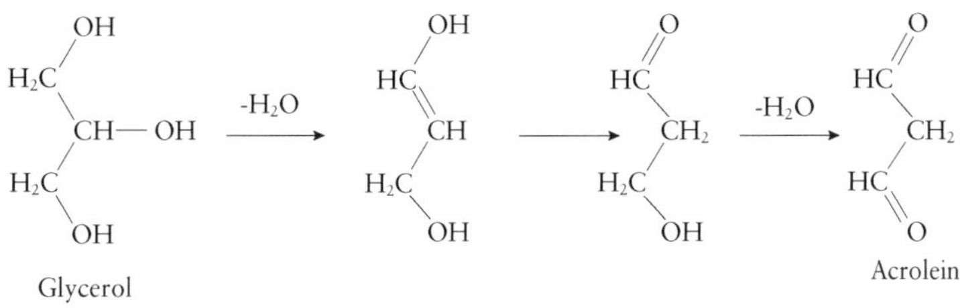

Scheme 4.34 Reaction mechanism for the catalytic dehydration of glycerol to acrolein

Selective transesterification and esterification are efficient methods to convert glycerol into monoglycerides and polyglycerides. Monoglycerides are, in general, obtained by the: i) hydrolysis of triglycerides; ii) glycerolysis of triglycerides (i.e., transesterification of glycerol with fatty methyl or ethyl esters, a base-catalysed transesterification of triglycerides with glycerol at elevated temperature); or iii) direct esterification of glycerol with fatty acids [138, 139]. All these structures can be used for the synthesis of highly branched polymers. Combination of fatty acids and glycerol has been exploited to synthesise amphiphilic polymers, which displayed self-assembly abilities in water [155].

Glycerol carbonate is widely used as a solvent for plastics and resins such as cellulose acetate, nylon, nitrocellulose and polyacrylonitrile. It has also become an interesting material in the chemical industry, mainly due to its role in the production of PC and PU [156]. Conversion of glycerol to glycerol carbonate is another economically significant alternative route to produce glycidol, which is an important source of new hyperbranched macromolecules, dendrimers and polyglycidol [157].

All evidence based on the current situation suggests that glycerol production will increase further. Its application as a feedstock for the synthesis of polymers (as replacement of some classical petrochemical routes) is crucial for increasing sustainability.

References

1. J.V. Crivello and R. Narayan, *Chemistry of Materials*, 1992, **4**, 692.

2. J.V. Crivello, R. Narayan and S.S. Sternstein, *Journal of Applied Polymer Science*, 1997, **64**, 2073.

3. J.V. Crivello and S. Kong, *Macromolecules*, 2000, **33**, 833.

4. A. Gandini in *Epoxy Polymers: New Materials and Innovation*, Eds., J.P. Pascault and R.J.J. Williams, Wiley-VCH, Weiheim, Germany, 2010.

5. M.S. Kim, K.W. Lee, T. Endo and S.B. Lee, *Macromolecules*, 2004, **37**, 5830.

6. S.J. Park, F.L. Jin and J.R. Lee, *Macromolecular Chemistry and Physics*, 2004, **205**, 2048.

7. F.L. Jin and S.J. Park, *Polymer International*, 2008, **57**, 577.

8. F.L. Jin and S.J. Park, *Materials Science and Engineering*, 2008, **A478**, 402.

9. G. Lligadas, J.C. Ronda, M. Galià, U. Biermann and J.O. Metzger, *Journal of Polymer Science, Part A: Polymer Chemistry Edition*, 2006, **44**, 634.

10. Z. Liu, K.M. Doll and R.A. Holser, *Green Chemistry*, 2009, **11**, 1774.

11. E. Del Río, M. Galià, V. Cádiz and G. Lligadas, *Journal of Polymer Science, Part A: Polymer Chemistry Edition*, 2010, **48**, 4995.

12. X.-Y. Jian, X.-P. An, Y.-D. Li, J.-H. Chen, M. Wang, J.-B. Zeng, *Macromolecules* 2015, **50**, 5729.

13. G. López Téllez, E. Vigueras-Santiago, S. Hernández-López and B. Bilyeu, *Designed Monomers & Polymers*, 2008, **11**, 435.

14. P. Czub, *Polymers for Advanced Technologies*, 2009, **20**, 194.

15. P. Czub, *Macromolecular Symposia*, 2009, **277**, 162.

16. J.D. Espinoza-Perez, B.A. Nerenz, D.M. Haagenson, Z.G. Chen, C.A. Ulven and D.P. Wiesenborn, *Polymer Composites*, 2011, **32**, 1806.

17. Z. Wang, X. Zhang, R. Wang, H. Kang, B. Qiao, J. Ma, L. Zhang and H. Wang, *Macromolecules*, 2012, **45**, 9010.

18. G. Lligadas, J.C. Ronda, M. Galià and V. Cádiz, *Journal of Polymer Science, Part A: Polymer Chemistry Edition*, 2006, **44**, 5630 and 6717.

19. N. Bouquillon and C. Fringant, *Polymer*, 2000, **41**, 8603.

20. H. Miyagawa, M. Misra, L.T. Drzal and A.K. Moharty, *Polymer Engineering & Science*, 2005, **45**, 487.

21. H. Miyagawa, A. Moharty, L.T. Drzal and M. Mizra, *Industrial & Engineering Chemistry Research*, 2004, **43**, 7001.

22. T. Takahashi, K. Hirayama, N. Teramoto and M. Shibata, *Journal of Applied Polymer Science*, 2008, **108**, 1596.

23. E. Doğan and S. Küsefoğlu, *Journal of Applied Polymer Science*, 2008, **110**, 1129.

24. F.I. Altuna, L.H. Espósito, R.A. Ruseckaite and P.M. Stefani, *Journal of Applied Polymer Science*, 2011, **120**, 789.

25. H. Pelletier, N. Belgacem and A. Gandini, *Journal of Applied Polymer Science*, 2006, **99**, 3218.

26. H. Pelletier and A. Gandini, *European Journal of Lipid Science and Technology*, 2006, **108**, 411.

27. S.N. Khot, J.J. Lascala, E. Can, S.S. Morye, G.I. Williams and G.R. Palmese, *Journal of Applied Polymer Science*, 2001, **82**, 703.

28. J. La Scala and R.P. Wool, *Polymer*, 2005, **46**, 61.

29. J. Lu, S. Khot and R.P. Wool, *Polymer*, 2005, **46**, 71.

30. R. Wang and T.P. Schumann, *eXPRESS Polymer Letters*, 2013, **7**, 272.

31. D.P. Pfister, Y. Xia and R.C. Larock, *ChemSusChem*, 2011, **4**, 703.

32. P. Furtwengler, L. Avérous, *Polymer Chemistry*, 2018, **9**, 4258–4287. L. Maisonneuve, G. Chollet, E. Grau, H. Cramail, *Oilseeds & fats Crops and Lipids*, 2016, **23**, D508. P. Alagi, S. C. Hong, *Macromolecular Research*, 2015, **23**, 1079.

33. Z.S. Petrovic, L.T. Yang, A. Zlatanic, W. Zhang and I. Javni, *Journal of Applied Polymer Science*, 2007, **105**, 2717.

34. S.D. Miao, S.P. Zhang, Z.G. Su and P. Wang, *Journal of Polymer Science, Part A: Polymer Chemistry Edition*, 2010, **48**, 243.

35. P. Rojek and A. Prociak, *Journal of Applied Polymer Science*, 2012, **125**, 2936.

36. H. Pawlik and A. Prociak, *Journal of Polymers and the Environment*, 2012, **20**, 438.

37. M. Jalilian, H. Yeganeh and M.N. Haghighi, *Polymers for Advanced Technologies*, 2010, **21**, 118.

38. M. Jalilian, H. Yeganeh and M.N. Haghighi, *Polymer International*, 2008, **57**, 1385.

39. L. Hojabri, X.H. Kong and S.S. Narine, *Biomacromolecules*, 2009, **10**, 884.

40. X. Kong and S.S. Narine, *Biomacromolecules*, 2007, **8**, 2203.

41. Y.S. Lu and R.C. Larock, *Biomacromolecules*, 2008, **9**, 3332.

42. Y.S. Lu and R.C. Larock, *Progress in Organic Coatings*, 2020, **69**, 31.

43. Y. Lu and R.C. Larock, *ChemSusChem*, 2010, **3**, 329.

44. Y. Lu and R.C. Larock, *ACS Symposium Series*, 2010, **1043**, 87.

45. T.E. Garrison, A.M. Kessler and R.C. Larock, *Polymer*, 2014, **55**, 1004.

46. S. Miao, P. Yong, Z. Su and S. Zhang, *Acta Biomaterialia*, 2014, **10**, 1692.

47. G. Cayl and S. Kusefoglu, *Journal of Applied Polymer Science*, 2010, **116**, 2433.

48. G. Cayl and S. Kusefoglu, *Journal of Applied Polymer Science*, 2008, **109**, 2948.

49. A.S. More, T. Lebarbe, L. Maisonneuve, B. Gadenne, C. Alfos and H. Cramail, *European Polymer Journal*, 2013, **49**, 823.

50. L. Hojabri, X. Kong and S.S. Narine, *Journal of Polymer Science, Part A: Polymer Chemistry Edition*, 2010, **48**, 3302.

51. L. Hojabri, X. Kong and S. S. Narine, *Biomacromolecules*, 2010, **11**, 911.

52. S.Y. Omprakash and S.P. Zoran, *ACS Symposium Series*, 2010, **1061**, 29.

53. A. Saralegi, L. Rueda, B. Fernandez-d'Arlas, I. Mondragon, A. Eceiza and M.A. Corcuera, *Polymer International*, 2012, **62**, 106.

54. C. Bueno-Ferrer, E. Hablot, M.D.C. Garrigos, S. Bocchini, L. Averous and A. Jimenez, *Polymer Degradation and Stability*, 2012, **97**, 1964.

55. C. Bueno-Ferrer, E. Hablot, F. Perrin-Sarazin, M.C. Garrigos, A. Jimenez and L. Averous, *Macromolecular Materials and Engineering*, 2012, **297**, 777.

56. D.V. Palaskar, A. Boyer, E. Cloutet, J-F. Le Meins, B. Gadenne, C. Alfos, C. Farcet and H. Cramail, *Journal of Polymer Science, Part A: Polymer Chemistry Edition*, 2012, **50**, 1766.

57. A. Boyer, C.E. Lingome, O. Condassamy, M. Schappacher, S. Moebs-Sanchez, Y. Queneau, B. Gadenne, C. Alfos and H. Cramail, *Polymer Chemistry*, 2012, **4**, 296.

58. A.F. Souza, A. Gandini, A.J.D. Silvestre, C. Pascoal Neto, J.J.C. Cruz Pinto, C. Eckerman and B. Holmbom, *Journal of Polymer Science, Part A: Polymer Chemistry Edition*, 2011, **49**, 2281.

59. A.F. Souza, A.J.D. Silvestre, A. Gandini and C. Pascoal Neto, *High Performance Polymers*, 2012, **24**, 4.

60. D. Quinzler and S. Mecking, *Angewandte Chemie International Edition*, 2010, **49**, 4306.

61. F. Stempfle, D. Quinzler, I. Heckler and S. Mecking, *Macromolecules*, 2011, **44**, 4159.

62. P. Roesle, C.J. Durr, H.M. Moller, L. Cavallo, L. Caporaso and S. Mecking, *Journal of the American Chemical Society*, 2012, **134**, 17696.

63. M.R.L. Furst, R.L. Goff, D. Quinzler, S. Mecking, C.H. Botting and D.J. Cole-Hamilton, *Green Chemistry*, 2012, **14**, 472.

64. H. Ebata, K. Toshima and S. Matsumura, *Macromolecular Bioscience*, 2008, **8**, 38.

65. K.S. Bisht, L.A. Henderson, R.A. Gross, D.L. Kaplan and G. Swift, *Macromolecules*, 1997, **30**, 2705.

66. M. Letizia Focarete, M. Scandola, A. Kumar and R.A. Gross, *Journal of Polymer Science, Part B: Polymer Physics Edition*, 2001, **39**, 1721.

67. M. de Geus, I. van der Meulen, B. Goderis, K. van Hecke, M. Dorschu, H. van der Werff, C.E. Koning and A. Heise, *Polymer Chemistry*, 2010, **1**, 525.

68. J. Cai, C. Liu, M. Cai, J. Zhu, F. Zuo, B.S. Hsiao and R.A. Gross, *Polymer*, 2010, **51**, 1088.

69. I. van der Meulen, E. Gubbels, S. Huijser, R.L. Sablong, C.E. Koning, A. Heise and R. Duchateau, *Macromolecules*, 2011, **44**, 4301.

70. M.P.F. Pepels, M. Bouyahyi, A. Heise and R. Duchateau, *Macromolecules*, 2013, **46**, 4324.

71. M. Bouyahyi, M.P.F. Pepels, A. Heise and R. Duchateau, *Macromolecules*, 2012, **45**, 3356.

72. W. Lu, J.E. Ness, W. Xie, X. Zhang, J. Minshull and R.A. Gross, *Journal of the American Chemical Society*, 2010, **132**, 15451.

73. N. Kolb and M.A.R. Meier, *Green Chemistry*, 2012, **14**, 2429.

74. Y. Yang, W. Lu, J. Cai, Y. Hou, S. Ouyang, W. Xie and R.A. Gross, *Macromolecules*, 2011, **44**, 1977.

75. T. Tsujimoto, H. Uyama and S. Kobayashi, *Biomacromolecules*, 2000, **2**, 29.

76. H. Uyama, M. Kuwabara, T. Tsujimoto and S. Kobayashi, *Biomacromolecules*, 2003, **4**, 211.

77. Y. Yang, W. Lu, X. Zhang, W. Xie, M. Cai and R.A. Gross, *Biomacromolecules*, 2009, **11**, 259.

78. J.E. White, J.D. Earls, J.W. Sherman, L.C. Lopez and M.L. Dettloff, *Polymer*, 2007, **48**, 3990.

79. S. Miao, S. Zhang, Z. Su and P. Wang, *Journal of Polymer Science, Part A: Polymer Chemistry Edition*, 2008, **46**, 4243.

80. Z. You, H. Cao, J. Gao, P.H. Shin, B.W. Day and Y. Wang, *Biomaterials*, 2010, **31**, 3129.

81. Y-R. Zhang, S. Spinella, W. Xie, J. Cai, Y. Yang, Y-Z. Wang and R.A. Gross, *European Polymer Journal*, 2013, **49**, 793.

82. P.J. Roumanet, F. Laflèche, N. Jarroux, Y. Raoul, S. Claude and P. Guégan, *European Polymer Journal*, 2013, **49**, 813.

83. E. Hablot, B. Donnio, M. Bouquey and L. Averous, *Polymer*, 2010, **51**, 5895.

84. E. Hablot, A. Tisserand, M. Bouquey and L. Averous, *Polymer Degradation and Stability*, 2011, **96**, 1097.

85. S. Samanta, J. He, S. Selvakumar, J. Lattimer, C. Ulven, M. Sibi, J. Bahr and B.J. Chisholm, *Polymer*, 2012, **54**, 1141.

86. J. He, S. Samanta, S. Selvakumar, J. Lattimer, C. Ulven, M. Sibi, J. Bahr and B.J. Chisholm, *Green Materials*, 2013, **1**, 114.

87. G. Walther, J. Deutsch, A. Martin, F-E. Baumann, D. Fridag, R. Franke and A. Kockritz, *ChemSusChem*, 2011, **4**, 1052.

88. N. Kolb, R. Hofsäβ and M.A.R. Meier, *European Journal of Lipid Science and Technology*, 2013, **115**, 729.

89. M. Winkler, M. Steinbis and M.A.R. Meier, *European Journal of Lipid Science and Technology*, 2014, **116**, 44.

90. R. Vanderhenst and S.A. Miller, *Green Materials*, 2013, **1**, 64.

91. C. Vilela, R. Rua, A.J.D. Silvestre and A. Gandini, *Industrial Crops and Products*, 2010, **32**, 97.

92. K.L. Beers and K. Matyjaszewski, *Journal of Macromolecular Science, Part A: Pure and Applied Chemistry*, 2001, **38**, 731.

93. T. Biedroń and P. Kubisa, *Macromolecular Rapid Communications*, 2001, **22**, 1237.

94. L. Lienafa, S. Monge and J-J. Robin, *European Polymer Journal*, 2009, **45**, 1845.

95. J.F.J. Coelho, E.Y. Carvalho, D.S. Marques, A.V. Popov, P.M. Goncalves and M.H. Gil, *Macromolecular Chemistry and Physics*, 2007, **208**, 1218.

96. F. Dutertre, P-Y. Pennarun, O. Colombani and E. Nicol, *European Polymer Journal*, 2011, **47**, 343.

97. S. Qin, J. Saget, J. Pyun, S. Jia and T. Kowalewski, *Macromolecules*, 2003, **36**, 8969.

98. G. Street, D. Illsley and S.J. Holder, *Journal of Polymer Science, Part A: Polymer Chemistry Edition*, 2005, **43**, 1129.

99. X. Zhu, Y. Gu, G. Chen, Z. Cheng and J. Lu, *Journal of Applied Polymer Science*, 2004, **93**, 1539.

100. K. Karaky, G. Clisson, G. Reiter and L. Billon, *Macromolecular Chemistry and Physics*, 2008, **209**, 715.

101. W. Xu, X. Zhu, Z. Cheng and J. Chen, *Journal of Applied Polymer Science*, 2003, **90**, 1117.

102. V. Raghunadh, D. Baskaran and S. Sivaram, *Polymer*, 2004, **45**, 3149.

103. Y. Xu, H. Becker, J. Yuan, M. Burkhardt, Y. Zhang, A. Walther, S. Bolisetty, M. Ballauff and A.H.E. Muller, *Macromolecular Chemistry and Physics*, 2007, **208**, 1666.

104. D.P. Chatterjee and B.M. Mandal, *Polymer*, 2006, **47**, 1812.

105. D.P. Chatterjee and B.M. Mandal, *Macromolecular Symposia*, 2006, **240**, 224.

106. D.P. Chatterjee and B.M. Mandal, *Macromolecules*, 2006, **39**, 9192.

107. J. Zhou, L. Wang, X. Dong, T. Chen, Q. Yang, Q. Tan and J. Wang, *European Polymer Journal*, 2007, **43**, 2088.

108. J. Zhou, L. Wang, X. Dong, Q. Yang, J. Wang, H. Yu and X. Chen, *European Polymer Journal*, 2007, **43**, 1736.

109. W. Jakubowski, J-F. Lutz, S. Slomkowski and K. Matyjaszewski, *Journal of Polymer Science, Part A: Polymer Chemistry Edition*, 2005, **43**, 1498.

110. G. Cayli and M.A.R. Meier, *European Journal of Lipid Science and Technology*, 2008, **110**, 853.

111. P.J. Roth, D. Kessler, R. Zentel and P. Theato, *Journal of Polymer Science, Part A: Polymer Chemistry Edition*, 2009, **47**, 3118.

112. M. Demetriou and T. Krasia-Christoforou, *Journal of Polymer Science, Part A: Polymer Chemistry Edition*, 2008, **46**, 5442.

113. L. Maisonneuve, T. Lebarbé, E. Grau and H. Cramail, *Polymer Chemistry*, 2013, **4**, 5472.

114. S. Rengasami and V. Mannari, *Progress in Organic Coatings*, 2013, **76**, 78.

115. B. Vazquez, S. Deb, W. Bonfield and J.S. Roman, *Journal of Biomedical Materials Research*, 2002, **63**, 88.

116. J.J. La Scala, J.M. Sands, J.A. Orlicki, E.J. Robinette and G.R. Palmese, *Polymer*, 2004, **45**, 7729.

117. C. Vilela and A. Gandini, Private Communication.

118. S. Kobayashi and H. Uyama, *Journal of Polymer Science, Part A: Polymer Chemistry Edition*, 2002, **40**, 192.

119. R. Hoogenboom, *Angewandte Chemie International Edition*, 2009, **48**, 7978.

120. N. Adams and U.S. Schubert, *Angewandte Chemie International Edition*, 2007, **48**, 7978.

121. A.M. Kelly and F. Wiesbrock, *Macromolecular Rapid Communications*, 2012, **33**, 1632.

122. R. Hoogenboom, *European Journal of Lipid Science and Technology*, 2011, **113**, 59.

123. B. Guillerm, S. Monge, V. Lapinte and J-J. Robin, *Macromolecular Rapid Communications*, 2012, **33**, 1600.

124. H. Mutlu and M.A.R. Meier, *European Journal of Lipid Science and Technology*, 2010, **112**, 10.

125. V.B. Veronese, R.K. Menger, M.M.D. Forte and C.L. Petzhold, *Journal of Applied Polymer Science*, 2011, **120**, 530.

126. D.G. Hayes, V.K. Mannam, R. Ye, M. Zhao, S. Ortega and M.C. Montiel, *Polymers*, 2012, **4**, 1037.

127. Y. Xu, Z. Petrovic, S. Das and G.L. Wilkes, *Polymer*, 2008, **49**, 4248.

128. G. Totaro, L. Cruciani, M. Vannini, G. Mazzola, D. Di Gioia, A. Celli and L. Sisti, *European Polymer Journal*, 2014, **56**, 174.

129. L. Maisonneuve, A.S. More, S. Foltran, C. Alfos, F. Robert, Y. Landais, T. Tassaing, E. Grau and H. Cramail, *RSC Advances*, 2014, **4**, 25795.

130. *The Future of Glycerol*, Eds., M. Pagliaro and Rossi, RSC Publishing, Cambridge, UK, 2008.

131. L. Prati, P. Spontoni and A. Gaiassi, *Topic in Catalalysis*, 2009, **52**, 288.

132. C. Crotti , E. Farnetti and N. Guidolin, *Green Chemistry*, 2010, **12**, 2225.

133. M. Ionescu and Z. S. Petrović, *European Journal of Lipid Science and Technology*, 2018, **120**, 1800004. A. Hejna, P. Kosmela, K. Formela, L. Piszczyk, J. T. Haponiuk, *Renewable and Sustainable Evergy Reviews*, 2016, **66**, 449. R. Ciriminna, B. Katryniok, S. Paul, F. Dumeignil, M. Pagliaro, *Organic Process, Research and Development*, 2015, **19**, 748. A. Martin and M. Richter, *European Journal of Lipid Science and Technology*, 2011, **113**, 100.

134. S. Salehpour and M.A. Dubé, *Macromolar Reaction Engineering*, 2012, **6**, 85.

135. S. Salehpour and M.A. Dubé, *Macromolecular Chemistry and Physics*, 2011, **212**, 1284.

136. H. Kimura, *Journal of Polymer Science, Part A: Polymer Chemistry Edition*, 2000, **38**, 664.

137. H. Kimura, *Polymers for Advanced Technology*, 2001, **12**, 697.

138. M. Pagliaro, R. Ciriminna, H. Kimura, M. Rossi and C. Della Pina, *Angewandte Chemie International Edition*, 2007, **46**, 4434.

139. C.H.C Zhou, J.N. Beltramini, Y-X Fan and G.Q.M. Lu, *Chemical Society Reviews*, 2008, **37**, 527.

140. A. Macierzanka and H. Szela, *Industrial & Engineering Chemistry Research*, 2004, **43**, 7744.

141. J.L. Lynn Jr., in *Bailey's Industrial Oil and Fat Products*, John Wiley & Sons, Hoboken, NJ, USA, 2005.

142. D.T. Johnson and K.A. Taconi, *Environment Progress*, 2009, **26**, 338.

143. J.R. Weiser, P.N. Zawaneh and D. Putnan, *Biomacromolecules*, 2011, **12**, 977.

144. M. Helou, J-M. Brusson, J-F. Carpentier and S.M. Guillaume, *Polymer Chemistry*, 2011, **2**, 2789.

145. P.N. Zawaneh, A.M. Doody, A.N. Zelikin and D. Putnam, *Biomacromolecules*, 2006, **7**, 3245.

146. A.N. Zelikin and D. Putnam, *Macromolecules*, 2005, **38**, 5532.

147. L-S. Wang, S-X. Cheng and R-X. Zhuo, *Macromolecular Rapid Communications*, 2004, **25**, 959.

148. T. Fukuoka, H. Habe, D. Kitamoto and K. Sakaki, *Journal of Oleo Science*, 2011, **60**, 369.

149. P.G. Parzuchowski, M. Jaroch, M. Tryznowski and G. Rokicki, *Macromolecules*, 2008, **41**, 3859.

150. P.G. Parzuchowski, M. Grabowska, M. Jaroch and M. Kusznerczuk, *Journal of Polymer Science, Part A: Polymer Chemistry Edition*, 2009, **47**, 3860.

151. X. Zhao, L. Liu, H. Dai, C. Ma, X. Tan and R. Yu, *Journal of Applied Polymer Science*, 2009, **113**, 3376.

152. B. Katryniok, S. Paul, V. Bellière-Baca, P. Rey and F. Dumeignil, *Green Chemistry*, 2010, **12**, 2079.

153. A. Corma, G.W. Huber, L. Sauvanaud and P. O'Connor, *Journal of Catalysis*, 2008, **257**, 163.

154. P.D. Pham, S. Monge, V. Lapinte, Y. Raoul and J.J. Robin, *European Journal of Lipid Science and Technology*, 2013, **115**, 28.

155. J. Rotta, P.D. Pham, V. Lapinte, R. Borsali, E. Minatti and J-J. Robin, *Macromolecular Chemistry and Physics*, 2014, **215**, 131.

156. V. Plasman and T. Caulier, *Plastics, Additives and Compounding*, 2005, **7**, 30.

157. R. Tokar, P. Kubisa and S. Penczek, *Macromolecules*, 1994, **27**, 320.

5 Metathesis Reactions Applied to Plant Oils and Polymers Derived from the Ensuing Products

5.1 General Considerations

Olefin metathesis ('metathesis' from the Greek 'change of position, transposition') [1] is a key reaction in organic synthesis because it allows preparation of molecules that are crucial to promote advances in medicine, biology and materials science. The importance of this reaction was demonstrated by the award of the 2005 Nobel Prize in chemistry to Yves Chauvin, Robert H. Grubbs, and Richard R. Schrock for elucidation of the reaction mechanisms and discovery of various highly efficient and selective catalysts related to metathesis [2].

Briefly, in their essence, olefin metathesis reactions induce reorganisation of the carbon atoms of two C=C double bonds to generate two new ones (**Scheme 5.1**). As stated by Hoveyda and Zhugralin in a very informative review published by *Nature* in 2007 [1], olefin metatheses may be classified as 'cross-metathesis' (CM), 'ring-closing metathesis' (RCM) or 'ring-opening metathesis' (ROM) [1, 3]. Metathesis reactions applied in polymer synthesis ('ring-opening metathesis polymerisation' (ROMP)) and acyclic diene metathesis (ADMET) polymerisation will be discussed in more detail later in this chapter.

In CM (**Scheme 5.1a**), $C_1=C_2$ and $C_3=C_4$ can be transposed into $C_1=C_3$ and $C_2=C_4$. In this case, if the reaction occurs with two $C_1=C_2$-type molecules, it is called 'self-metathesis' (SM) (**Scheme 5.1b**). In RCM (the most widely used process), two terminal alkenes react with the catalyst to generate a cyclic olefin to release a smaller olefin ($C_2=C_4$ in **Scheme 5.1c**). Finally, in ROM (**Scheme 5.1d**), a cyclic olefin reacts with a linear (acyclic) olefin to generate an acyclic diene.

The reaction itself was discovered by Karl Ziegler in the 1950s while working in what would later become known as Ziegler–Natta catalysis. During his studies on ethylene polymerisation, he noticed that addition of specific metal-based catalysts led to formation of 1-butene instead of a saturated long-chain hydrocarbon. However, the generally accepted mechanism of the metathesis reaction was not proposed until 1971 by Chauvin [4], who demonstrated that it proceeds *via* a [2+2] cycloaddition to form a metallocyclobutane intermediate followed by a reversible cycloelimination which generates transalkylidenated products (**Scheme 5.2**). [5]

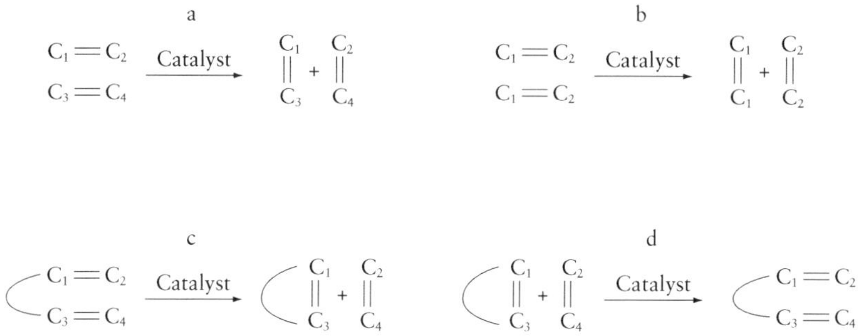

Scheme 5.1 a) Cross-metathesis; b) self-metathesis; c) ring-closing metathesis; and d) ring-opening metathesis [1]

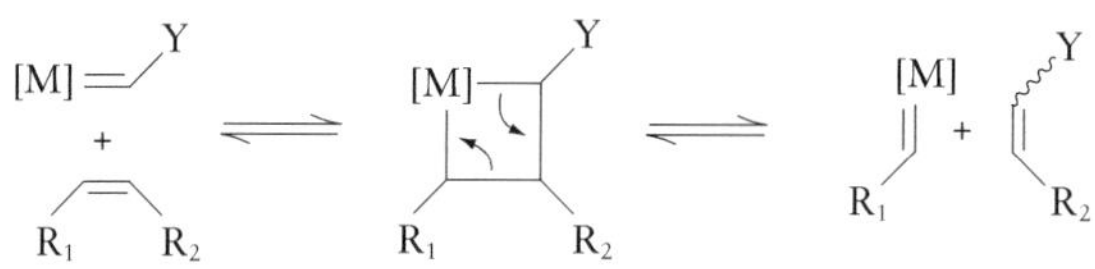

Scheme 5.2 Representative mechanism of the metathesis reaction for olefin [4, 5]

The thermodynamic equilibrium of the reaction is an important factor in terms of achieving a higher control of the process to drive the reaction towards the desired products. This equilibrium is governed primarily by Le Chatelier's principle. One of the products is removed continuously from the reaction medium to shift the equilibrium progressively toward formation of products. This strategy is especially effective in the case of CM reactions involving terminal olefins because the volatile ethylene gas formed as the byproduct can be removed easily and if high-molecular-weight polymers are sought (as in the case of ADMET reactions). Indeed, the possibility of having such easy control is a major qualitative advantage and a differential over other polymerisation mechanisms. Also, for ROM and ROMP reactions, the energy released during the ring-opening of cyclic structures is often sufficient to drive them forward.

Despite some inherent singularities, what is common to all metathesis reactions is the great importance attributed to the catalyst employed, which determines the efficiency and efficacy of the entire process. Some of the catalysts that are commonly used in olefin metathesis bear, in most cases, a molybdenum [6] or a ruthenium [7] atom because Mo=C or Ru=C double bonds act as points of contact between the catalyst and olefins. Complexes with other metals (e.g., tungsten, rhenium, osmium) promote olefin metathesis but these exhibit lower stability and/or reactivity, and have not been investigated as extensively [1].

Only development of ruthenium-based catalysts has allowed tolerance to a singular functional group to operate without requiring experiments to be conducted under an inert atmosphere [5]. They can be used with substrates that carry an alcohol, a carboxylic acid, or an aldehyde, but can be rendered inactive in the presence of structurally exposed amines and phosphines. The reverse is true for molybdenum-based catalysts, but they are generally more active than ruthenium catalysts, and demand preparation and handling under an inert atmosphere [1]. **Table 5.1** is a summary of the functional-group tolerance of some general catalysts for olefin metathesis [8].

Table 5.1 Functional-group tolerance of early and late transition metal olefin metathesis catalysts

Titanium	Tungsten	Molybdenum	Ruthenium	
Acids	Acids	Acids	*Olefins*	↑ Increasing reactivity
Alcohols, water	Alcohols, water	Alcohols, water	Acids	
Aldehydes	Aldehydes	Aldehydes	Alcohols, water	
Ketones	Ketones	*Olefins*	Aldehydes	
Esters, amides	*Olefins*	Ketones	Ketones	
Olefins	Esters, amides	Esters, amides	Esters, amides	

Adapted from T.M. Trnka and R.H. Grubbs, *Accounts of Chemical Research*, 2001, 34, 18 [8]

Titanium- and tungsten-based catalysts tend to react with olefinated ketones and esters, whereas molybdenum-based counterparts are more reactive toward olefins (though they can also react with aldehydes and other polar or protic groups). Ruthenium-based catalysts react preferentially with C=C double bonds over most other moieties, a feature that makes them unusually stable toward alcohols, amides, aldehydes, and carboxylic acids. Because of this trend, the functional-group tolerance of an olefin metathesis catalyst can be increased by focusing on a later class of transition metal such as ruthenium [8].

Scheme 5.3 illustrates the most commonly used ruthenium-based olefin metathesis catalysts: (I) the first well-defined, metathesis-active ruthenium alkylidene complex

[5, 9]; (II) first-generation Grubbs' catalyst; (III) second-generation Grubbs' catalyst; and (IV) second-generation Hoveyda–Grubbs' catalyst (HG-II) [5].

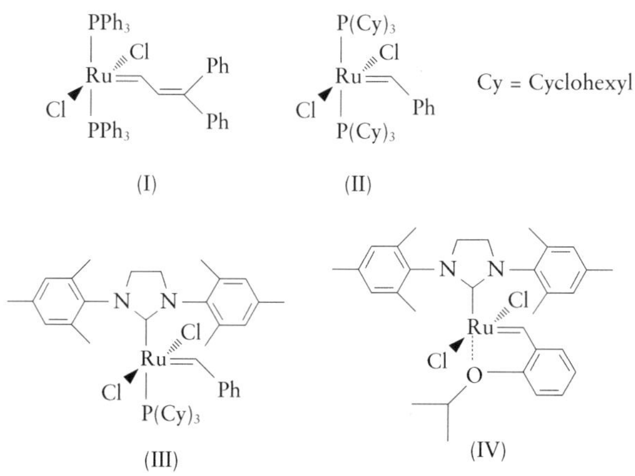

Scheme 5.3 Most commonly used ruthenium-based olefin metathesis catalysts [5]

Metathesis reactions applied to vegetable oils have been used extensively as a versatile tool for the synthesis of monomers and polymers, mainly because most of the fatty-acids chains that constitute them contain at least one C=C double bond. Therefore, this characteristic of vegetable oils (allied to their cheap prices, high availability and renewable character) and the undeniable potential of the metathesis reactions in organic synthesis offer a range of possibilities to prepare unsaturated molecules that were previously unavailable, difficult to obtain by any other means, or available only through low-yield processes [10–12].

Evidence of the unquestionable potential of this association is the tremendous increase in the number of studies dedicated to the subject as well as exploiting this strategy in large-scale processes. For example, Elevance Renewable Sciences® recently announced the commercial availability of Inherent™ C18 Diacid (their trade name for octadecanedioic diacid), which is the first commodity based on Elevance's proprietary metathesis technology [13]. In this case, long-chain bifunctional monomers are produced through metathesis and give rise to a wide range of possibilities for production of predominantly aliphatic polymers.

This chapter describes the most relevant advances related to application of metathesis reactions to modification of vegetable oils and their fatty-acid derivatives, as well as use of these reactions to prepare novel macromolecular structures and materials.

5.2 Metathesis Reactions as Tools for the Synthesis of Monomers and Polymers Derived from Vegetable Oils

As mentioned above, the metathesis reaction is a very useful tool in general organic synthesis. Within the scenario of vegetable oils and their derivatives for enabling the synthesis of monomers and their polymerisation, it has become indispensable because it opens possibilities for creating new molecules and materials from a worldwide set of available renewable resources.

As a tool for monomer synthesis, the metathesis reaction is applied in fatty-acid modifications. That is, exploitation of the reactions shown in **Scheme 5.1**: CM, SM, RCM and ROM. However, due to the inherent long, acyclic, aliphatic structure of vegetable oils and their derivatives, and in view of further applications in polymerisation (or any other modification reactions), most studies on application of metathesis systems to unsaturated fatty acids have been based on SM and CM mechanisms (**Scheme 5.1a** and **b**) to synthesise the desired ensuing molecules (usually monomers).

Once the molecule (modified or not) bears the appropriate structure, it can be polymerised directly *via* metathesis. Then, the possibilities consist mostly in applying ADMET polymerisation or ROMP. ADMET polymerisation, subject of most of the studies on polymerisations *via* metathesis, is generally classified as a step-growth polycondensation-type reaction that produces strictly linear chains from unconjugated dienes. ROMP involves a chain-growth-type polymerisation, and is generally limited by the availability of suitable monomers.

The following sections of this chapter describe in more detail (but not exhaustively) the fundamentals and state-of-the-art of such processes keeping in mind that this subject is being updated constantly by enthusiastic researchers all over the world.

5.2.1 Metathesis Reactions for Monomer Synthesis

As mentioned above, most studies that employ metathesis for monomer synthesis from vegetable-oil derivatives take advantage of SM and CM mechanisms (**Scheme 5.4**) rather than RCM and ROM.

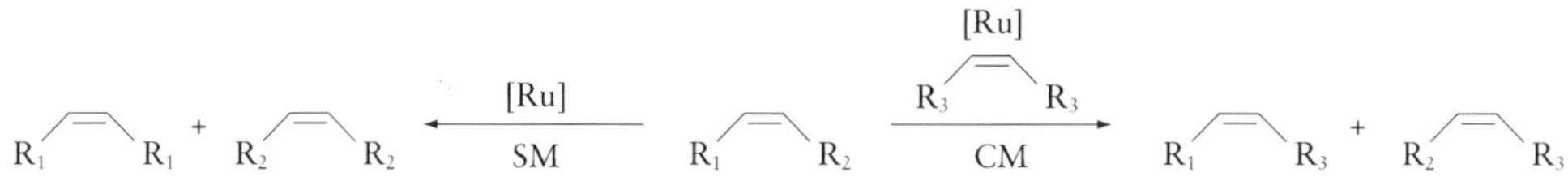

Scheme 5.4 Self-metathesis and cross-metathesis reactions

The general representation depicted in **Scheme 5.4** ignores the possibility of CM with a species containing two different groups next to the unsaturation (R_3-=-R_4 instead of R_3-=-R_3). This is also possible, but usually, if this is the case, four different molecules can be formed as products. This phenomenon leads to lower yields of the desired product, and requires more careful purification steps.

In SM and CM, combinations that lead to ethylene formation are often preferred (i.e., if the structures involved contain terminal unsaturations). This phenomenon is because easy removal of ethylene from the reaction medium contributes very effectively to achieve full conversion and/or a higher yield, compared with the cases where this situation is not possible.

To illustrate the recently discovered pathways to functional monomers, Meier and colleagues studied the synthesis of a long-chain diester from a ω-hydroxy fatty acid derived from palmitic acid. The idea was to transform the ω-hydroxyl function into a mesylate, followed by an elimination reaction to prepare the ω-unsaturated fatty acid methyl ester (FAME), which was dimerised by a SM coupling to obtain the desired C_{30} diester (**Scheme 5.5**) [14]. This macromonomer was then polymerised with diols and diamines to prepare long-chain polyesters and polyamides (PA) with interesting thermal properties, such as a melting temperature (T_m) of 109 °C for the polyester and 166 °C for the PA.

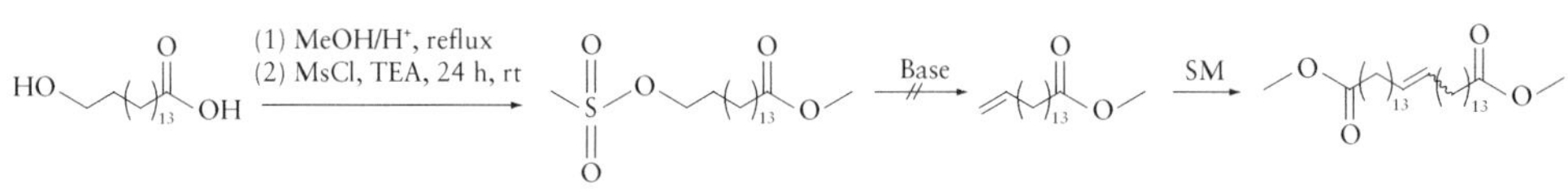

Scheme 5.5 Synthesis of a C_{30} α,ω-diester derived from ω-hydroxy palmitic acid. TEA: Triethylamine [14]

Stempfle and co-workers also used SM to synthesise the dimethyl-1,19-nonadecanedioate monomer from the methyl ester of high oleic sunflower oil. After hydrogenation and reduction to the corresponding diol, a simple polycondensation reaction catalysed by titanium alkoxides allowed formation of a polymer of relatively high molecular weight [number average molecular weight (Mn) ≈30 kDa], which was injected and presented an elongation at break of >600% and a Young's Modulus of 400 MPa [15].

Novel PA were prepared by polycondensation between diacids and diamines catalysed by 1,5,7-triazabicyclo[4.4.0]dec-5-ene (TBD) in a very interesting study described by Meier and Mutlu [16]. In this study, unsaturated C20 chain PA with a T_m 180–226 °C were synthesised by polycondensation of the SM product of 10-undecenoic acid and various diamines. Fradet and colleagues studied various diamines (aliphatic, cycloaliphatic, benzylic) and used the same methodology [17, 18].

CM is also used in monomer synthesis from vegetable oils and their fatty acids. The increasing number of publications related to this subject over the last decade indicates its importance in the preparation of new molecules derived from renewable resources.

In 2005, Patel and co-workers [19] undertook a set of CM reactions with sunflower, canola, linseed and soya oils and 2-butene using three ruthenium-based catalysts. Each of the vegetable oils gave excellent conversion to butenolysis products and high catalyst turnovers. Higher turnovers were obtained for oils containing predominantly oleic acid (e.g., sunflower oil) compared with those with higher proportions of polyunsaturated fatty acids (e.g., linseed oil).

A much more recent approach [20] described the CM of methyl 10-undecenoate with dimethyl maleate (**Scheme 5.6**) using a wide range of ruthenium-based catalysts from 0.1 to 1.0 mol%. Optimisation of the process led to nearly quantitative yields of the desired diester, which has excellent potential to be applied in further polymerisations.

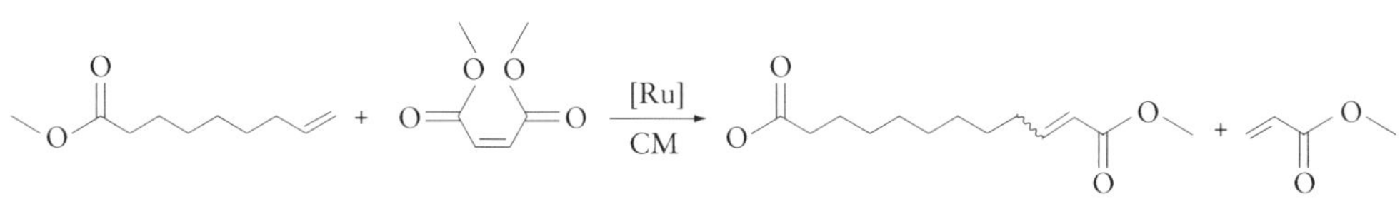

Scheme 5.6 Cross-metathesis of methyl 10-undencenoate and dimethyl maleate [20]

Some interesting investigations illustrate the use of CM for the synthesis of novel PA. Spiccia and co-workers [21] described a multi-catalytic sequence to produce methyl

11-aminoundecanoate, a Nylon 11 precursor. This structure was obtained by a ruthenium-catalysed CM and a highly regioselective palladium-catalysed amination–hydrogenation reaction from the canola-oil derivative methyl undec-9-enoate. This study adopted an approach similar to that reported by Meier and co-workers involving the ruthenium-catalysed CM of 1,4-diacetoxybut-2-ene with methyl undec-10-enoate [22]. Meier and co-workers had also reported the CM of fatty-acid derivatives with allyl chloride as potential precursors to the synthesis of polyesters and PA [23].

Miao and co-workers described the ruthenium-catalysed CM of the unsaturated fatty-acid derivative 10-undecenenitrile prepared from castor oil with methyl acrylate to produce a C12 nitrile ester with a high turnover number. The linear C12 α,ω-amino ester (an important PA precursor) could be prepared by the direct tandem CM/hydrogenation in the presence of *t*-BuOK (**Scheme 5.7**) [24].

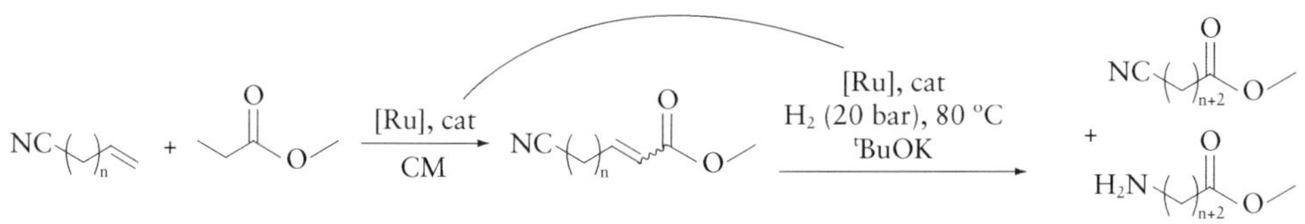

Scheme 5.7 Synthetic route to amino esters by sequential metathesis and hydrogenation reactions [18, 24]

A beautiful study by Winkler and Meier [25] reported on the bulk olefin CM reaction of benzyl carbamates derived from unsaturated FAME prepared *via* a catalytic Lossen rearrangement procedure with methyl acrylate. High yields were achieved with low catalyst loadings (0.5 mol%). At ambient temperature and atmospheric hydrogen pressure, it was possible to cleave the carbamate function and the remaining double bond was hydrogenated in one step, leading to the amino FAME in quantitative yields. The corresponding PA displayed a high Young's Modulus (1,480 MPa for the C15 PA, while the commercially available Nylon 11 and 12 have values of 1,740 and 2,320 MPa), respectively (**Scheme 5.8**).

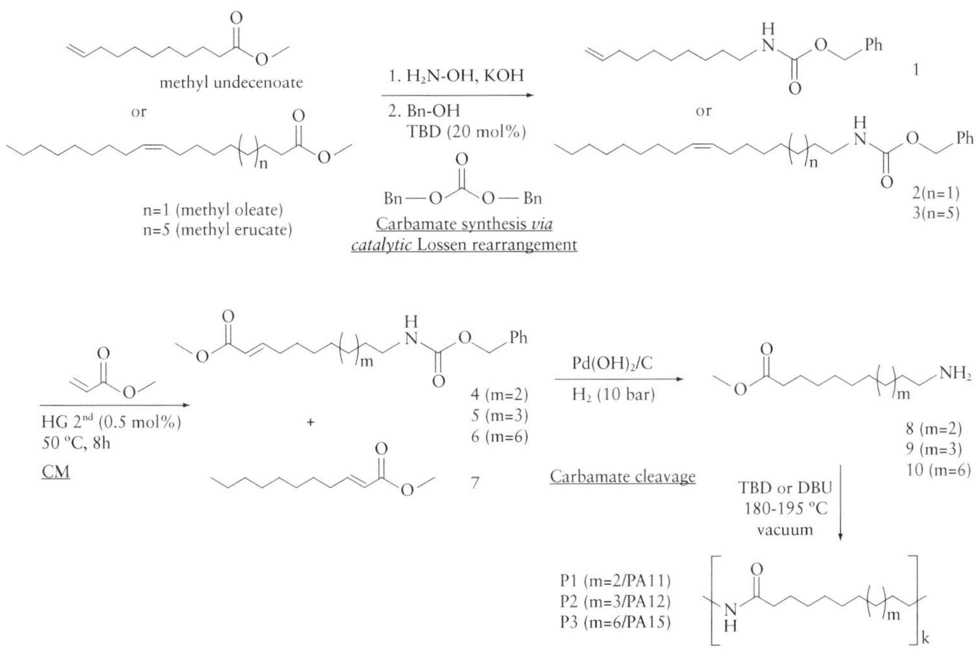

Scheme 5.8 Synthesis of renewable AB-type monomers *via* the Lossen rearrangement, cross-metathesis, and subsequent deprotection. DBU: 1,8-Diazabicyclo[5.4.0]undec-7-ene [25]

CM is also very useful if the objective is to modify branched polymers or to introduce some convenient functional groups. This was the case of two original studies conducted by the Meier's research team. The first study [26] described the reaction of a poly(2-oxazoline) containing terminal double bonds in the side chains with various functional acrylates in the presence of a HG-II. The second (more recent) study [27] involved the synthesis of a polymalonate bearing C9 aliphatic side chains with terminal double bonds used for grafting-onto reactions by ruthenium-catalysed CM.

The examples mentioned above are just a few of the studies that have been reported but not discussed here. They illustrate how wide the possibilities are of creating new molecules derived from a renewable resource using metathesis reactions. A more curious reader can be rewarded by looking for other important works that were regrettably omitted in this succinct summary.

5.2.2 Olefin Metathesis Applied to Polymer Synthesis

In general, the two most widely spread synthetic approaches to olefin metathesis polymerisations have been mentioned: ADMET polymerisations and ROMP. Each requires a different set of conditions to achieve success. The following sections deal with examples of both, and also some special cases that might be interesting.

5.2.2.1 Acyclic Diene Metathesis Polymerisation

The ADMET polymerisation reaction is a special type of olefin metathesis used to polymerise terminal dienes to polyenes (**Scheme 5.9**).

The first successful ADMET polymerisation was reported by Wagener and colleagues [28]. They polymerised 1,5-hexadiene and 1,9-decadiene to 1,4-polybutadiene [with a weight average molecular weight (Mw) of 28 kDa] and polyoctenylene (Mw = 108 kDa), respectively, using a tungsten-based catalyst that required extremely dry conditions to avoid side reactions. Recent advances in the development of very active and stable catalysts now allow the synthesis of various polymer architectures with relative ease.

An important feature of this reaction is that it is driven by the release of volatile ethylene gas. Consequently, though olefin metathesis with ruthenium-based catalysts is, in general, done under very mild conditions, ADMET polymerisation benefits greatly from conditions that promote the diffusion and expulsion of ethylene (i.e., higher reaction temperatures, application of vacuum, vigorous stirring) so that better conversion and high values of the molecular weight of the final polymer can be achieved.

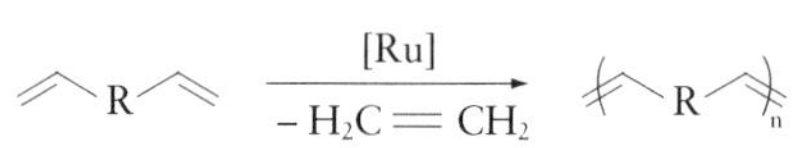

Scheme 5.9 The ADMET polymerisation reaction

As stated by Cramail, [18] a disadvantage of such metathesis reactions is the possibility of side reactions (e.g., isomerisation of the unsaturation) to create a flaw in the polymer structure and thus decrease its thermo-mechanical properties. This inherent characteristic of the process can be circumvented by optimising different parameters, such as the temperature or the nature of the catalyst, to suppress (or at least limit) double-bond isomerisation [28–34]. Also, the aliphatic polyesters obtained by metathesis show, in general, promising thermo-mechanical properties, but a low T_m. A possible reason for this drawback is the remaining internal double bonds in the final polymer structure, which tend to conspire against a high degree of crystallisation. To overcome this limitation, it is often useful to undertake hydrogenation of the final polymer [35–38].

Most of the studies in the last decade have concentrated on the ADMET polymerisation of fatty acids and its derivatives to obtain polyesters as the final materials. In 2008, Meier and Rybak [39] described the synthesis of an α,ω-diene by transesterification of 10-undecenoic acid with the corresponding alcohol (10-undecenol). ADMET polymerisation of this monomer was conducted to achieve a C20 unsaturated polyester.

Cramail and co-workers [40] described the synthesis of two linear and branched α,ω-diene monomers from castor and vernonia oils to develop linear low-density polyethylene and very low-density polyethylene-like polyesters. Methyl 10-undecenoate and 12-hydroxystearic acid, obtained from ricinoleic acid, were employed for the synthesis of the branched α,ω-diene, and a second monomer of similar structure (but linear) was synthesised from methyl 10-undecenoate and 1,12-dodecanediol (obtained from vernolic acid). Further ADMET copolymerisation of the dienes led to a series of copolyesters with Mn values of 14–62 kDa, T_m of 13.2–87.4 °C, and with various degrees of branching depending on the feed ratio between the two α,ω-dienes.

Another interesting work on the ADMET polymerisation of a castor-oil-based diene, 1,3-di-10-undecenoxy-2-propanol, was conducted by the research team of Cádiz [41]. Authors also utilised 10-undecenol as a renewable comonomer to functionalise polymer chains and react them with 4,4'-methylene diphenyl diisocyanate (MDI) to yield a series of polyurethanes (PU).

Kreye, Tóth and Meier used ADMET polymerisation extensively to synthetise novel (co)polymers from hydroxyl-bearing unsaturated compounds to prepare polyesters containing functional groups in their side chains. Those materials were based entirely on structures obtained from: rapeseed (i.e., oligomers from a ferulic-acid derivative); polymers from derivatives of oleic acid and erucic acid; copolymers from derivatives of oleic acid and ferulic acid; derivatives of erucic acid and ferulic acid [42]. Another hydroxyl-bearing α,ω-diene was synthesised by the reaction of two equivalents of 10-undecenoic acid with one equivalent of 1,3-dichloro-2-propanol yielding, through ADMET polymerisation, a semi-crystalline polyester with a Mn of ≈5 kDa, a glass transition temperature (T_g) of –25 °C and a T_m of 41 °C [18, 43].

Although less frequently than with polyesters, it is possible to produce other polymers from ADMET polymerisation. Mecking and co-workers [44] described the synthesis of polyacetals and polycarbonates (PC) with a sparse and systematically varied density of functional groups generated by ADMET copolymerisation of unfunctionalised undeca-1,10-diene with bis(undec-10-en-1-yloxy)methane or di(undec-10-en-1-yl) carbonate, followed by exhaustive hydrogenation (**Scheme 5.10**).

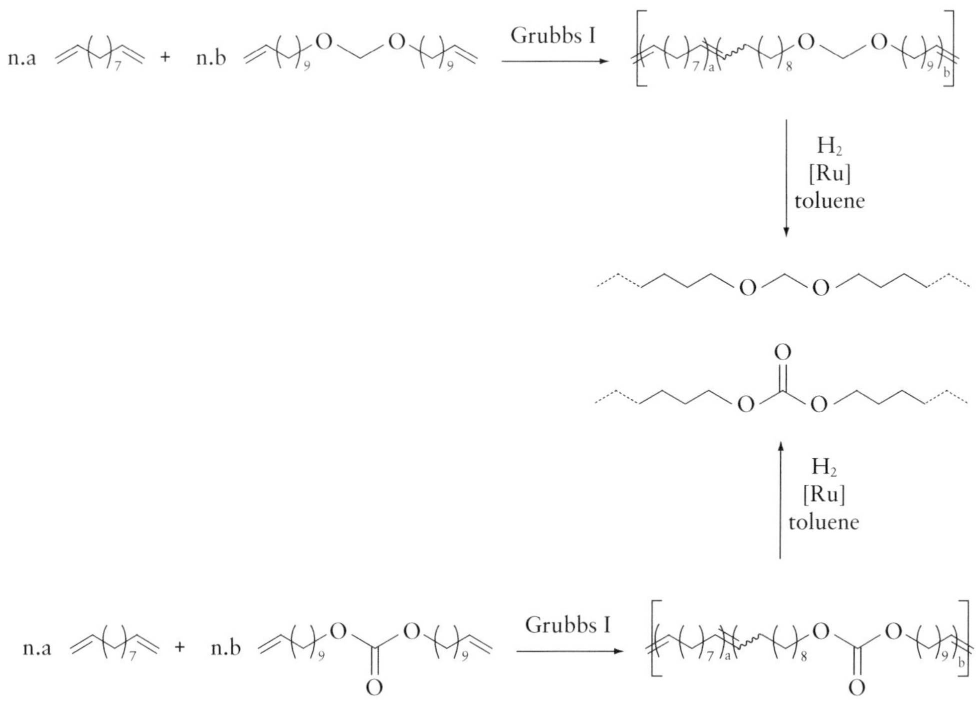

Scheme 5.10 Long-spaced polyacetals (upper scheme) and polycarbonates (lower scheme) by ADMET copolymerisation and subsequent hydrogenation [44]

Also for the synthesis of renewable PC, Meier and co-workers [45] described dimethyl carbonate and 1,5,7-triazabicyclo[4.4.0]dec-5-ene (TBD) as catalysts for the synthesis of a carbonate derivative obtained by the reaction of 10-undecenol with dimethyl carbonate, which led to bis-unsaturated compounds that were polymerised by ADMET. In this case, high molecular weights were obtained (Mn = 10–28 kDa).

Recently, renewable PU were synthesised from vegetable-oil derivatives by the research team of Cramail [46] in a new and very interesting approach to obtain an α,ω-diene urethane monomer by converting 10-undecenoic acid into the corresponding acyl azide, followed by urethanisation with 10-undecenol (**Scheme 5.11**).

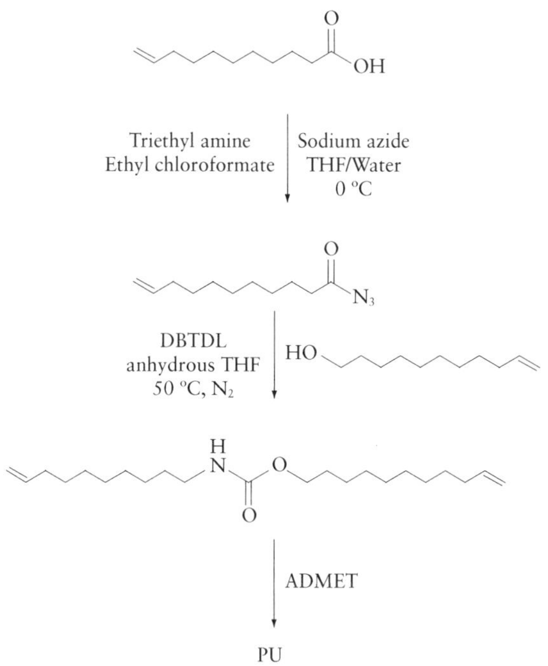

Scheme 5.11 Structure of the monomer used to prepare novel renewable polyurethanes. DBTDL: Dibutyltin dilaurate and THF: tetrahydrofuran [47]

PA were prepared *via* ADMET polymerisation [16] in parallel to another approach described above for monomer synthesis *via* SM reactions. Long-chain aliphatic α,ω-dienes with two symmetrically spaced amide segments were polymerised *via* ADMET polymerisation in this investigation as (**Scheme 5.12**). Long-chain polyethers were synthesised using a monomer derived from castor oil through its commercially available derivative of undecenoic acid, 11-bromo-1-undecene, *via* ADMET reactions to generate a material resembling polyethylene [47]. Polymers obtained from these processes exhibited Mn values of 8–16 kDa and T_m values of 40–60 °C.

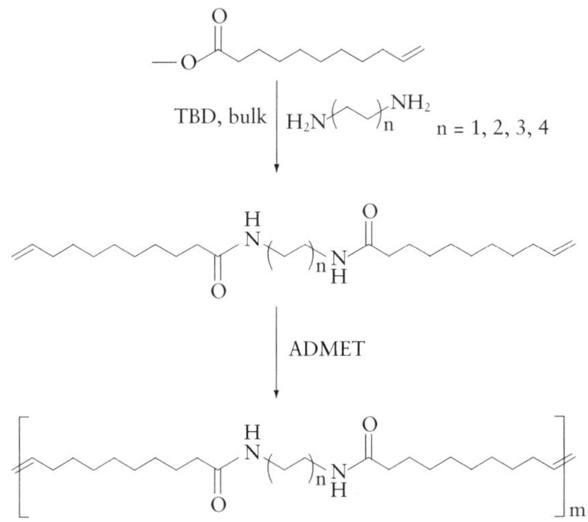

Scheme 5.12 ADMET route to prepare renewable polyamides [16]

Another significant application of the ADMET polymerisation relates to the preparation of star-shaped polymers, which are branched macromolecules in which several linear polymer chains are attached to a unique branching point or core [48]. Montero de Espinosa, Winkler and Meier [49] described an ADMET approach to obtain those architectures (three- and four-arm) using small tri-acrylates and tetra-acrylates. More recently, Unverferth and Meier [50] reported the synthesis of well-defined star-shaped polymers *via* a head-to-tail ADMET polymerisation whereby di(trimethylolpropane)tetra-acrylate (four-arm) and dipentaerythritol hexa-acrylate (six-arm) served as core units, and fatty acid-derived 10-undecenyl acrylate as asymmetric α,ω-diene monomers. In this case, star-shaped polymers containing arms of 10 or 20 monomer units with an α,ω-unsaturated ester backbone and their subsequent post-polymerisation *via* a base-catalysed Thia–Michael addition were prepared.

The ADMET polymerisation reaction associated with vegetable oils and their derivatives is now a fundamental tool for the preparation of novel materials. Increasing interest in materials that are not derived from fossil resources is a driving force stimulating researchers to develop methods and hence to create new materials that can deflect our dependence on petroleum. This blooming strategy can, therefore, be expected to proceed further.

5.2.2.2 Acyclic Triene Metathesis Polymerisation

Acyclic triene metathesis (ATMET) polymerisation is a particular ADMET application useful in vegetable oils. ADMET can also be used for the polymerisation of triglycerides, offering the possibility to obtain branched polymers directly from plant oils without prior chemical modification, thereby helping to further minimise the environmental impact of such materials. This type of polymerisation was termed ATMET with reference to monomer functionality [51].

Despite its usefulness, there have been only a few investigations related to this subject. All were conducted between 2008 and 2011 and headed by Biermann, Meier and/or Metzger [52–55].

Pioneering work [52] described the ATMET reaction with a HG-II of a synthesised model triglyceride with different amounts of methyl acrylate as a chain stopper to prevent full crosslinking of the triglyceride, and to obtain branched polymers of different molecular weights in a straightforward one-step, one-pot approach (**Scheme 5.13**).

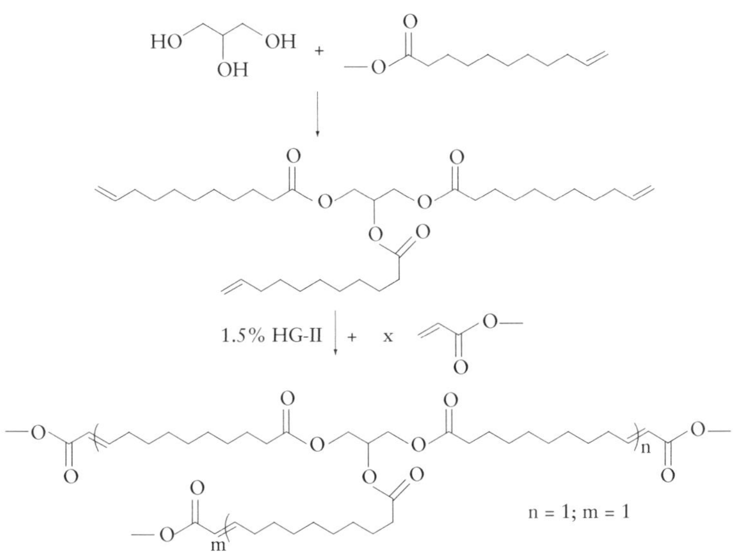

Scheme 5.13 Synthesis of branched macromolecules *via* ATMET polymerisation with a chain stopper [52]

Ensuing materials had Mn values of 3–5 kDa, indicating that the branching reaction had taken place. A more systematic study was conducted some years later [53] to isolate dimeric triglycerides obtained as minor products besides branched ATMET polymerisation. This first work, conducted with a model triglyceride, was followed by direct application of high oleic sunflower oil (i.e., macromonomers in which the internal double bonds replaced terminal ones) [54] in a more detailed study focusing on characterisation of the obtained oligomers and polymers, as well as isolation and characterisation of some low-molecular-weight intermediates of ATMET reactions (Mn ≈1–5 kDa).

This work involved the synthesis of branched polyols *via* ATMET polymerisation starting from the same model triglyceride depicted in **Scheme 5.13**, and its reaction with MDI to produce shape-memory PU [55]. Mn values of those polyols were 1.3–3 kDa, and the PU derived from them exhibited good thermo-mechanical properties.

These four studies on ATMET polymerisation led to interesting macromolecules and oligomers, but only for polymers of low molecular weight. The nature of vegetable oils and the difficulty in isolating 100% pure triglycerides from them could be an impediment in development of this method.

5.2.2.3 Ring-opening Metathesis Polymerisation

ROMP is a type of olefin metathesis chain-growth polymerisation (**Scheme 5.14**).

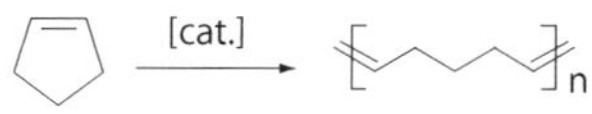

Scheme 5.14 ROMP reaction of cyclopentene (schematic)

The driving force for this reaction is the relief of ring strain in cyclic olefins (e.g., norbornene or cyclopentene). If more than one type of strained, unsaturated ring is present in the reaction medium, a copolymer is formed. This phenomenon is the first limitation of ROMP reactions: the availability of this mandatory strained cyclic structure. Various backbones can be created through monomer functionalisation, but such alterations can negatively affect ring strain (and hence the success of the corresponding ROMP).

ROMP is not the most suitable method if fatty acids are used as feedstock because of the low ring strain of the ensuing unsaturated cyclic monomers. Thus, for vegetable oils to participate in ROMP, modification of fatty-acid chains is necessary, and this can be achieved by coupling fatty-acid derivatives with conventional unsaturated cyclic building blocks [18, 56].

This is the case of the first study reporting on the ROMP of fatty-acid derivatives [57], a subject that has been studied by the research team of Larock for almost a decade. In this first publication, castor oil was functionalised with 5-norbornene-2,3-dicarboxylic anhydride. Resulting monomer structure was submitted to ring-opening metathesis copolymerisation with cyclooctene (**Scheme 5.15**) to produce thermoset polyesters with good thermo-mechanical properties.

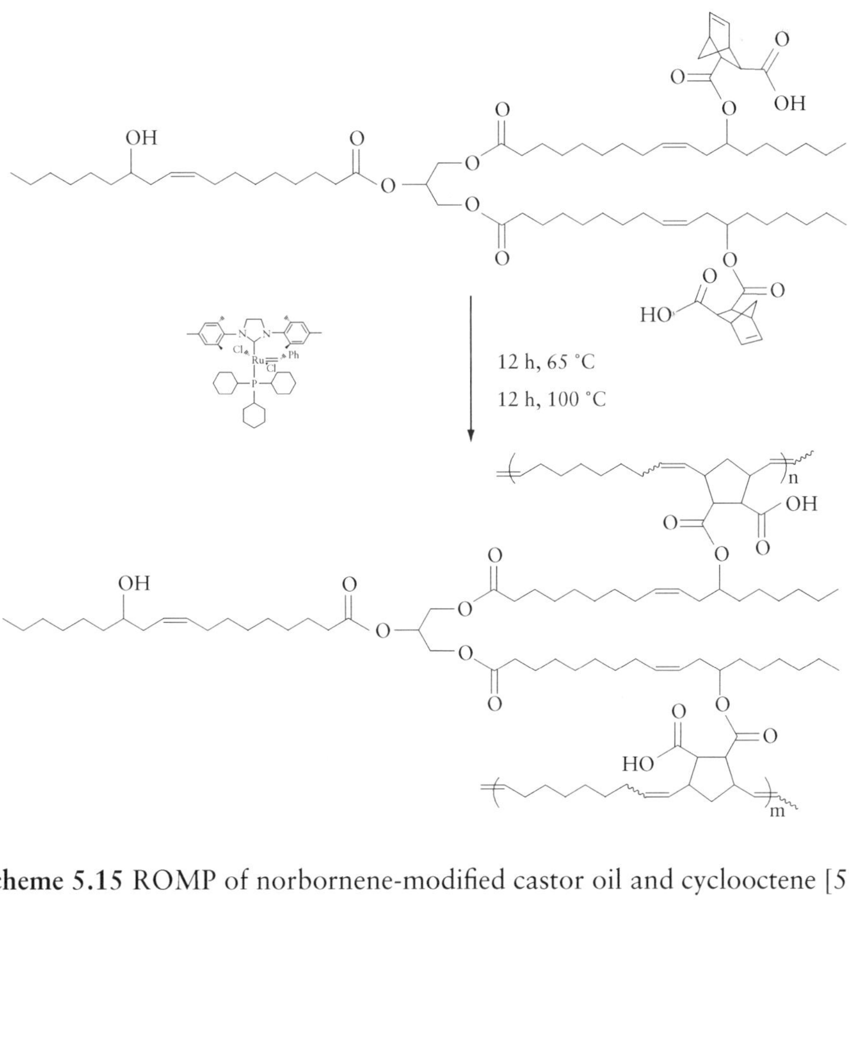

Scheme 5.15 ROMP of norbornene-modified castor oil and cyclooctene [57]

The Diels–Alder reaction of cyclopentadiene and linseed oil produces Dilulin, a chemical commodity commercialised by Cargill. Dilulin was used in ROMP with a norbornene-based agent to produce crosslinked materials [58]. The thermal stability and crosslink density of the resulting thermosets increased with increasing amounts of the crosslinking agent [56]. In the same year, the research team of Larock adopted the same approach to produce thermosets *via* the ROMP of Dilulin and dicyclopentadiene [59], which was used as a matrix for composites with glass fibre [60]. Analyses of scanning electron micrographs of surface morphology revealed weak adhesion between the fibre and polymer matrix. However, resins and the corresponding composites displayed an increase in the T_g, E', σ_{max}, E, and toughness as dicyclopentadiene content was increased.

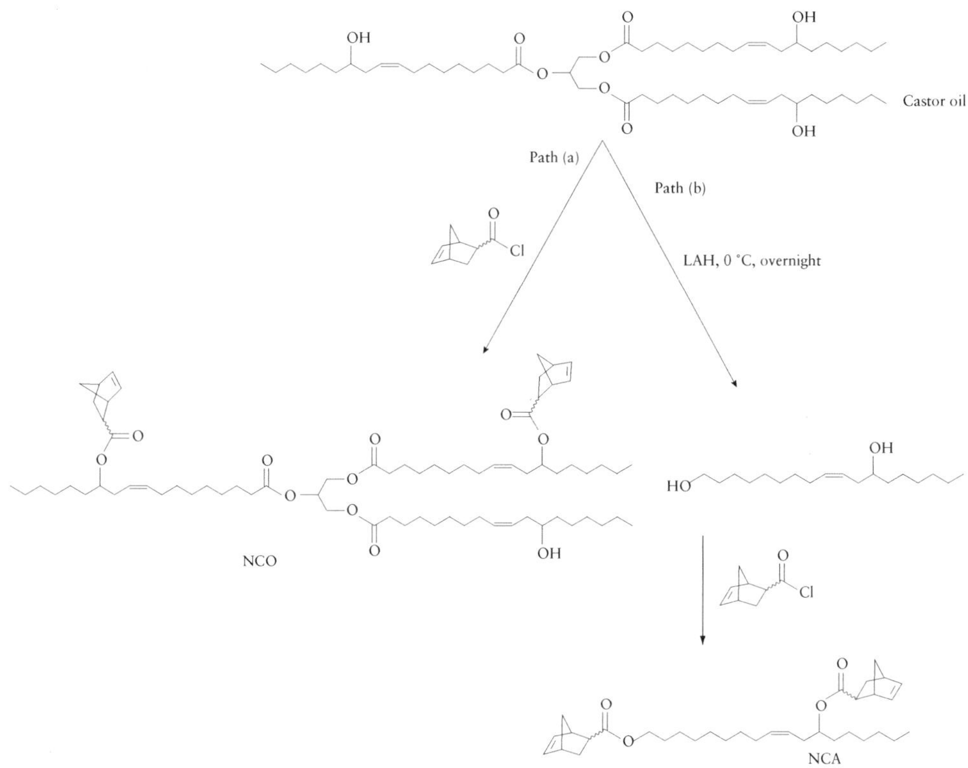

Scheme 5.16 Preparation of ROMP-susceptible monomers from castor oil. LAH: Lithium aluminum hydride [61]

In a subsequent study [61], castor oil was used in the synthesis of two monomers: norbornenyl-functionalised castor oil (NCO), and norbornenyl-functionalised castor oil alcohol (NCA) (**Scheme 5.16**). These monomers were submitted to ROMP with 0.125 wt% of a second-generation Grubbs' catalyst, and resulted in novel castor oil-based rubbery-to-rigid bio-renewable plastics with crosslink densities of 318–6,028 mol/m^3.

Last but not least, Meier and Mutlu reported on the synthesis of various ester-functionalised norbornenes using fatty acids of different chain lengths (C6 to C18) [62]. Their ROMP led to polymers with Mn values ≤165 kDa. Some properties of these materials were dependent upon the chain length of the fatty acid used to prepare them. For instance, the T_g decreased from 102 to –32 °C for polymers based on fatty acids with a chain length of 6–18 carbons, and the T_m decreased from 30 to 6 °C with the same increase in chain length [16].

Despite being an interesting tool for polymerisation if associated with vegetable oils, ROMP is very limited in scope, which is why there are so few reports addressing this subject.

5.2.2.4 Special Cases of Acetal Metathesis Polymerisation and Alternating Diene Metathesis Polymerisation

This section deals with two methods of metathesis polymerisation which, although much less wide-ranging in applications than the methods described above, are interesting tools for the preparation of polymers from vegetable-oil derivatives.

Acetal metathesis polymerisation (AMP) is an interesting tool because, besides involving a mechanism that allows for application of molecules derived from vegetable oils, it leads to the formation of *polyacetals*. Cellulose (the most abundant biopolymer on earth) has polymeric chains made of repeating units of glucose connected by acetal functional groups that hydrolyse readily under aqueous acidic conditions. Therefore, other polymers bearing acetal moieties are equally susceptible to hydrolytical cleavage, a property that allows the materials to be readily degradable in the environment in a relatively short timescale [18, 63].

The basic principle of this polymerisation reaction is based on the acetal interchange illustrated in **Scheme 5.17**.

Two studies have been published on utilisation of AMP for preparation of polymers from vegetable oils. Both studies used fatty acids for the synthesis of long-chain diols, which were submitted to preliminary acetal interchange using dimethoxymethane or diethoxymethane to give α,ω-diacetals monomers polymerisable *via* AMP.

The first work [64] described the synthesis of nonadecane-1,19-diol and tricosane-1,23-diol from oleic acid and erucic acid, respectively. For the synthesis of diacetals starting from diols and dimetoxymethane, use of 20 mol% methanesulfonic acid as a catalyst produced the desired 2,4,24,26-tetraoxaheptacosane from nonadecane-1,19-diol at 85% yield. Also, use of 20 mol% trifluoromethanesulfonic acid as a catalyst led to the formation of 2,4,28,30-tetraoxahentriacontane from tricosane-1,23-diol at 79% yield. Difunctional acetal monomers were then subjected to AMP at 80–100 °C, 2–4 mol% of *p*-toluenesulfonic acid as a catalyst, and reduced pressure to remove the dimethoxymethane byproduct. Authors obtained polymers with Mn values of 17 and 22 kDa, with the higher value arising from the longer-chain diol/diacetal.

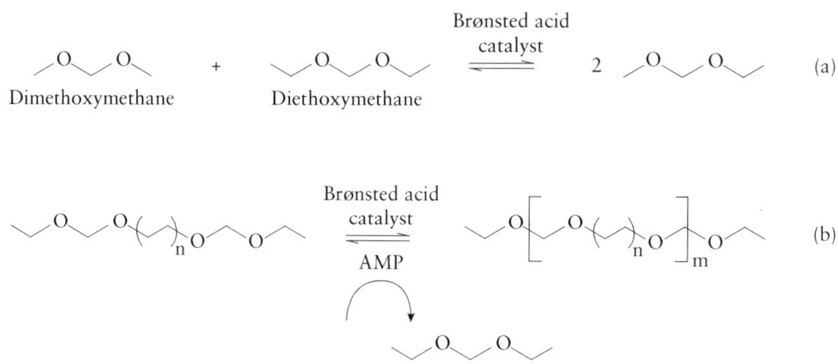

Scheme 5.17 Acetal interchange reaction a) and the same principle applied to polymerisation by the acetal metathesis polymerisation mechanism b)

The second work [63] employed the same AMP approach but instead used 1,10-decanediol (derivative of castor oil) as a starting material to arrive at a polyacetal with a molecular weight of 22 kDa and T_m of 59 °C.

Perfectly alternating copolymers can be obtained *via* alternating diene metathesis (ALTMET) polymerisation (**Scheme 5.18**), a concept introduced by Slugovc and colleagues more than a decade ago [65]. This concept is based on the high CM selectivity between terminal double bonds and acrylates in the presence of a second-generation Grubbs' catalyst. If a terminal diene (considered a type-I olefin) reacts with a diacrylate (classified as a type-II olefin that homodimerises much more slowly [66]) in the presence of a second-generation Grubbs' catalyst, the initial reversible ADMET polymerisation step of the terminal diene is followed by an irreversible insertion of the diacrylate. Moreover, the resulting substituted acrylates do not participate in secondary metathesis reactions [67]. The reaction of different terminal dienes and diacrylates led to AB copolymers with degrees of alternation ≤99%.

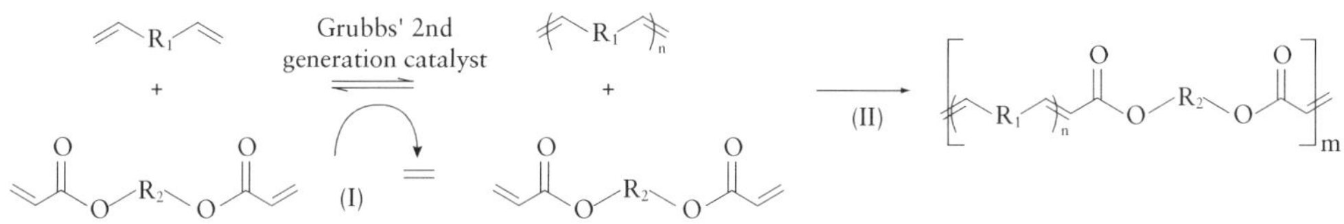

Scheme 5.18 Proposed mechanism for ALTMET polymerisation with the first step as a reversible ADMET reaction (I) and the second step an irreversible insertion of the diacrylate (II) [65]

The only study that correlates the ALTMET polymerisation reaction with a vegetable-oil derivative is quite recent [68]. It describes the preparation of alternating copolymers with functional phosphate groups. Diene I (**Scheme 5.19**) was synthesised from 10-undecenol, and its ALTMET polymerisation with the diacrylate II was carried out in sequence.

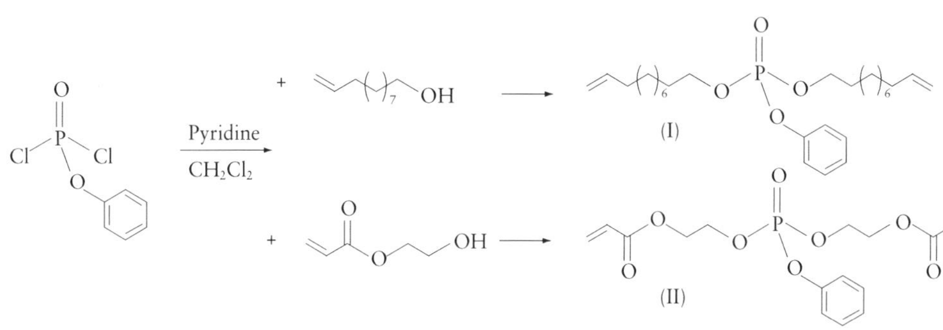

Scheme 5.19 Synthesis of a diene and a diacrylate to be copolymerised further to give AB-alternating copolymers by ALTMET polymerisation [68]

Authors stated that, using the ALTMET approach, molecular weights could be controlled by changing the relative stoichiometry of the diene/diacrylate. Using 0.95 equivalent of the diene and 1.0 equivalent of the diacrylate, a copolymer with Mn = 14 kDa, polydispersity index (PDI) = 1.89, and 97% AB alternation was obtained. However, a copolymer with Mn = 36 kDa, PDI = 1.95 and 94% AB alternation ensued by increasing the diene content to 1.05 equivalent. This difference corroborates the mechanism proposed by Slugovc [65]. That is, an ADMET homopolymerisation of the diene initially produces an unsaturated polymer scaffold to which subsequent insertions of the diacrylate generate the final AB-alternating structure because an excess of the diacrylate shortens the polymer chains, whereas an excess of the diene can give higher molecular weights as a result of the oligomeric blocks of polyphosphoesters [68].

AMP and ALTMET mechanisms are interesting (but very specific tools) that require very specific molecular structures to be applied successfully in polymer design. It follows that application of these concepts to oleochemistry is very limited, as shown by the paucity of studies dealing with polymers from vegetable oils.

References

1. A.H. Hoveyda and A.R. Zhugralin, *Nature*, 2007, **480**, 243.

2. *http://www.nobelprize.org/nobel_prizes/chemistry/laureates/2005/(March 2014).*

3. *Handbook of Metathesis*, Ed., R.H. Grubbs, Wiley-VCH, Weinheim, Germany, 2003.

4. L. Hérisson and Y. Chauvin, *Makromolekulare Chemie*, 1971, **141**, 161.

5. A. Rybak, P.A. Fokou and M.A.R. Meier, *European Journal of Lipid Science and Technology*, 2008, **110**, 797.

6. R.R. Schrock and A.H. Hoveyda, *Angewandte Chemie International Edition*, 2003, **42**, 4592.

7. S.T. Nguyen and T.M. Trnka in *Handbook of Metathesis*, Ed., R.H. Grubbs, Wiley-VCH, Weinheim, Germany, 2003.

8. T.M. Trnka and R.H. Grubbs, *Accounts of Chemical Research*, 2001, **34**, 18.

9. S.T. Nguyen, L.K. Johnson, R.H. Grubbs and J.W. Ziller, *Journal of the American Chemical Society*, 1992, **114**, 3974.

10. J.C. Mol, *Green Chemistry*, 2002, **4**, 5.

11. J.C. Mol, *Topics in Catalysis*, 2004, **27**, 97.

12. M.A.R. Meier, *Macromolecular Chemistry and Physics*, 2009, **210**, 1073.

13. Elevance Renewable Sciences, 2013. *http://www.elevance.com/platforms/engineered-polymers-and-coatings/inherent-c18-diacid.*

14. N. Kolb, M. Winkler, C. Syldatk and M.A.R. Meier, *European Polymer Journal*, 2014, **51**, 159.

15. F. Stempfle, B.S. Ritter, R. Mülhaupt and S. Mecking, *Green Chemistry*, 2014, **16**, 2008.

16. H. Mutlu and M.A.R. Meier, *Macromolecular Chemistry and Physics*, 2009, **210**, 1019.

17. F. Pardal, S. Salhi, B. Rousseau, M. Tessier, S. Claude and A. Fradet, *Macromolecular Chemistry and Physics*, 2008, **209**, 64.

18. L. Maisonneuve, T. Lebarbé, E. Grauab and H. Cramail, *Polymer Chemistry*, 2013, **4**, 5472.

19. J. Patel, J. Elaridi, W.R. Jackson, A.J. Robinson, A.K. Serelis and C. Such, *Chemical Communications*, 2005, **44**, 5546.

20. A. Behr, S. Toepell and S. Harmuth, *RSC Advances*, 2014, **4**, 16320.

21. N.D. Spiccia, E. Border, J. Illesinghe, W.R. Jackson and A.J. Robinson, *Synthesis*, 2013, **45**, 1683.

22. M. von Czapiewski, O. Kreye, H. Mutlu and M.A.R. Meier, *European Journal of Lipid Science and Technology*, 2013, **115**, 76.

23. T. Jacobs, A. Rybak and M.A.R. Meier, *Applied Catalysis A: General*, 2009, **353**, 32.

24. X. Miao, C. Fischmeister, P.H. Dixneuf, C. Bruneau, J.L. Dubois and J.L. Couturier, *Green Chemistry*, 2012, **14**, 2179.

25. M. Winkler and M.A.R. Meier, *Green Chemistry*, 2014, **16**, 3335.

26. L. Montero de Espinosa, K. Kempe, U.S. Schubert, R. Hoogenboom and M.A.R. Meier, *Macromolecular Rapid Communications*, 2012, **33**, 2023.

27. N. Kolb and M.A.R. Meier, *European Polymer Journal*, 2013, **49**, 843.

28. B. Schmidt, *European Journal of Organic Chemistry*, 2004, **9**, 1865.

29. S.E. Lehman, Jr., J.E. Schwendeman, P.M. O'Donnell and K.B. Wagener, *Inorganica Chimica Acta*, 2003, **345**, 190.

30. F.C. Courchay, J.C. Sworen and K.B. Wagener, *Macromolecules*, 2003, **36**, 8231.

31. M. Arisawa, Y. Terada, K. Takahashi, M. Nakagawa and A. Nishida, *The Chemical Record*, 2007, **7**, 238.

32. P.A. Fokou and M.A.R. Meier, *Journal of the American Chemical Society*, 2009, **131**, 1664.

33. S.H. Hong, D.P. Sanders, C.W. Lee and R.H. Grubbs, *Journal of the American Chemical Society*, 2005, **127**, 17160.

34. P.A. Fokou and M.A.R. Meier, *Macromolecular Rapid Communications*, 2010, **31**, 368.

35. J. Trzaskowski, D. Quinzler, C. Bährle and S. Mecking, *Macromolecular Rapid Communications*, 2011, **32**, 1352.

36. C. Vilela, A.J.D. Silvestre and M.A.R. Meier, *Macromolecular Chemistry and Physic*, 2012, **213**, 2220.

37. F. Stempfle, P. Ortmann and S. Mecking, *Macromolecular Rapid Communications*, 2013, **34**, 47.

38. H. Mutlu, R. Hofsa, R.E. Montenegro and M.A.R. Meier, *RSC Advances*, 2013, **3**, 4927.

39. A. Rybak and M.A.R. Meier, *ChemSusChem*, 2008, **1**, 542.

40. T. Lebarbé, M. Neqal, E. Grau, C. Alfos and H. Cramail, *Green Chemistry*, 2014, **16**, 1755.

41. E. del Rio, G. Lligadas, J.C. Ronda, M. Galià, M.A.R. Meier and V. Cádiz, *Journal of Polymer Science, Part A: Polymer Chemistry Edition*, 2011, **49**, 518.

42. O. Kreye, T. Tóth and M.A.R. Meier, *European Polymer Journal*, 2011, **47**, 1804.

43. L. Monteiro de Espinosa, M.A.R. Meier, J.C. Ronda, M. Galià and V. Cádiz, *Journal of Polymer Science, Part A: Polymer Chemistry Edition*, 2010, **48**, 1649.

44. P. Ortmann, I. Heckler and S. Mecking, *Green Chemistry*, 2014, **16**, 1816.

45. H. Mutlu, J. Ruiz, S.C. Solleder and M.A.R. Meier, *Green Chemistry*, 2012, **14**, 1728.

46. T. Lebarbé, A. Sudhakar More, P.S. Sane, E. Grau, C. Alfos and H. Cramail, *Macromolecular Rapid Communications*, 2014, **35**, 479.

47. O. Türünç, L. Montero de Espinosa and M.A.R. Meier, *Macromolecular Rapid Communications*, 2011, **32**, 1357.

48. K. Inoue, *Progress in Polymer Science*, 2000, **25**, 453.

49. L. Montero de Espinosa, M. Winkler and M.A.R. Meier, *Macromolecular Rapid Communications*, 2013, **34**, 1381.

50. M. Unverferth and M.A.R. Meier, *Polymer*, 2014, DOI 10.1016/j.polymer.2014.06.035.

51. *Green Polymerization Methods*, Eds., R.T. Mathers and M.A.R. Meier, Wiley-VCH, Weinheim, Germany, 2011.

52. P.A. Fokou and M.A.R. Meier, *Macromolecular Rapid Communications*, 2008, **29**, 1620.

53. U. Biermann, M.A.R. Meier, W. Butte and J.O. Metzger, *European Journal of Lipid Science and Technology*, 2011, **113**, 39.

54. U. Biermann, J.O. Metzger and M.A.R. Meier, *Macromolecular Chemistry and Physics*, 2010, **211**, 854.

55. E. del Río, G. Lligadas, J.C. Ronda, M. Galià, V. Cádiz and M.A.R. Meier, *Macromolecular Chemistry and Physics*, 2011, **212**, 1392.

56. R.L. Quirino and R.C. Larock in *Renewable and Sustainable Polymers*, ACS Symposium Series, American Chemical Society: Washington, DC, USA, 2011, **1063**, 37.

57. P.H. Henna and R.C. Larock, *Macromolecular Materials and Engineering*, 2007, **292**, 1201.

58. T.C. Mauldin, K. Haman, X. Sheng, P. Henna, R.C. Larock and M.R. Kessler, *Journal of Polymer Science, Part A: Polymer Chemistry Edition*, 2008, **46**, 6851.

59. P. Henna and R.C. Larock, *Journal of Applied Polymer Science*, 2009, **112**, 1788.

60. P.H. Henna, M.R. Kessler and R.C. Larock, *Macromolecular Materials and Engineering*, 2008, **293**, 979.

61. Y. Xia and R.C. Larock, *Polymer*, 2010, **51**, 2508.

62. H. Mutlu and M.A.R. Meier, *Journal of Polymer Science, Part A: Polymer Chemistry Edition*, 2010, **48**, 5899.

63. A.G. Pemba, J.A. Flores and S.A. Miller, *Green Chemistry*, 2013, **15**, 325.

64. S. Chikkali, F. Stempfle and S. Mecking, *Macromolecular Rapid Communications*, 2012, **33**, 1126.

65. S. Demel, C. Slugovc, F. Stelzer, K. Fodor-Csorba and G. Galli, *Macromolecular Rapid Communications*, 2003, **24**, 636.

66. A.K. Chatterjee, T-L. Choi, D.P. Sanders and R.H. Grubbs, *Journal of the American Chemical Society*, 2003, **125**, 11360.

67. H. Mutlu, L. Montero de Espinosa and M.A.R. Meier, *Chemical Society Reviews*, 2011, **40**, 1404.

68. L. Ding, X-Q. Zheng, R. Lu, J. An and J. Qiu, *Polymer International*, 2014, **63**, 997.

6 Thiol-ene and Thiol-yne Reactions for the Transformation of Oleochemicals into Monomers and Polymers

6.1 General Considerations

The *thiol-ene reaction*, which has been used for 110 years [1], corresponds to the simple hydrothiolation of a C=C double bond (**Scheme 6.1**). However, there are some peculiarities of this reaction that make it remarkably attractive today. It is a tool for researchers in all areas of organic synthesis due to recognition of its '*click*' characteristics.

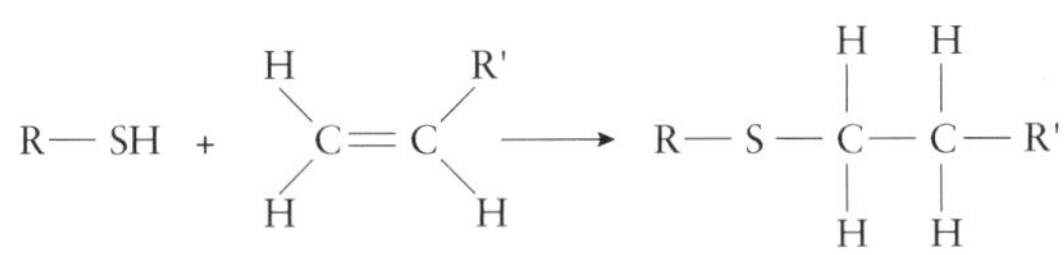

Scheme 6.1 General thiol-ene coupling reaction

'Click chemistry' is a concept developed by Sharpless and co-workers [2, 3] more than a decade ago. It describes a way of conducting organic reactions. It aims at *labelling* reactions that can produce a wide range of functional synthetic molecules and organic materials with high efficiency under various mild conditions.

Many definitions can be found in literature to describe the main principles that a reaction must follow to be considered as click. Hoyle and Bowman [4] summarised the main characteristics of modular click reactions as including: a) high yields with byproducts (if any) that are removable by non-chromatographic processes; b) regiospecificity and stereospecificity; c) insensitivity to oxygen or water; d) mild, solventless (or aqueous) reaction conditions; e) orthogonality with other common organic synthesis reactions; and f) amenability to a wide variety of readily available starting compounds.

As time progresses, this concept is becoming increasingly grounded in the mind of organic chemists, mainly because of the increasing worldwide 'green' philosophy that demands development of environmentally friendly molecules and strategies that move in the direction of more sustainable chemical processes. Therefore, because it was considered as click, the thiol-ene reaction has been the focus of many studies in the last decade. It has served as an extraordinary instrument for readily designed molecules (often monomers and polymers) with the additional advantage of its eco-friendly connotation. Today, thiol-ene click chemistry can be applied effectively to a large number of applications ranging from optical components and adhesives to high-impact energy-absorbing materials [4].

The reaction shown in **Scheme 6.1** occurs most of the time *via* free-radical addition to electron-rich/electron-poor C=C double bonds (this mechanism will be discussed in more detail below). However, in the case of addition to an α,ω-unsaturation (**Scheme 6.2**), the reaction needs mild base/nucleophilic catalysis, which is slightly less efficient than the radical-mediated mechanism because the C=C double bond is electron-deficient and must be activated [5].

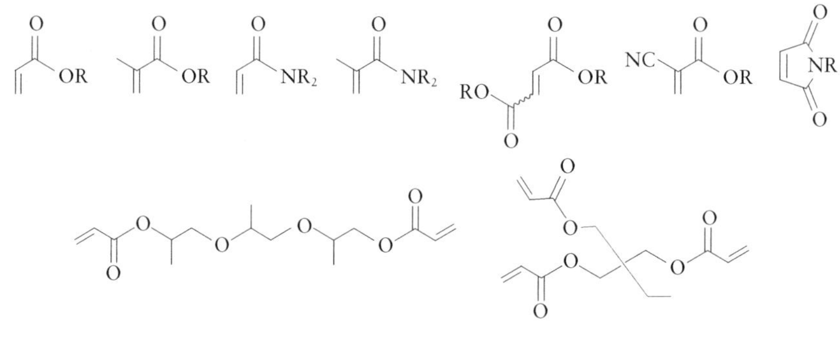

Scheme 6.2 Examples of substrates susceptible to hydrothiolation *via* a base/nucleophile-mediated process [5]

Under free-radical conditions (often induced by photochemical means), thiol-ene reactions proceed *via* a typical chain process with initiation, propagation and termination steps (**Scheme 6.3**). The mechanism involves a thiyl radical *cis–trans* isomerisation, with concurrent formation of a thiol-ene product. Initiation requires treatment of a thiol with a photo-initiator under irradiation, and results in formation of a thiyl radical (R–S•). Propagation involves direct addition of the thiyl radical across the C=C double bond to yield an intermediate carbon-centred radical, followed by chain transfer to a second molecule of thiol to give the thiol-ene addition product with an anti-Markovnikov orientation, and concomitant generation of a new thiyl radical. Possible termination reactions occur through typical radical–radical coupling processes [5].

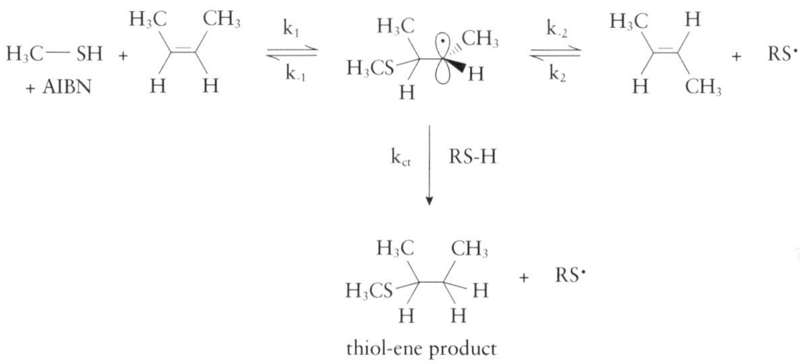

Scheme 6.3 Thiol-ene free-radical mechanism. AIBN: Azobisisobutyronitrile [5]

Addition of the thiol across the ene double bond is exothermic, with reaction enthalpies ranging from –10.5 kcal mol^{-1} for the electron-rich vinyl ether double bond to –22.6 kcal mol^{-1} for the electron-poor double bond of, for example, a *N*-alkyl maleimide [4].

One of the main advantages of the thiol-ene reaction is that, in principle, any thiol can be employed, including highly functional species. Reactivity will be a function of the strength of the S–H bond and of the cleavage mechanism (homolytic or heterolytic lysis). Also, such reactions are, in general, extremely rapid, and can be complete in seconds, often at ambient temperature and pressure. They are tolerant to oxygen and moisture, and proceed with near-quantitative formation of the desired product in a regioselective fashion [5].

Successful association between oleochemistry and thiol-ene reactions is demonstration of the immense potential of sustainable processes. A renewable resource as widely available as a vegetable oil, with the exceptional number of possibilities of molecules that can be obtained from it, together with a method as simple and efficient as the thiol-ene reaction, demonstrate that this strategy is much more than a tendency but instead a reality in polymer chemistry.

This chapter does not intend to be exhaustive, but aims instead at illustrating the impressive versatility and potential of the thiol-ene click reactions associated with fatty acids and their derivatives, as well as to highlight examples demonstrating the broad utility of this particular combination in monomer and polymer synthesis.

6.2 Thiol-ene Reaction as a Tool for the Synthesis of Monomers and Polymers Derived from Vegetable Oils

Associated with vegetable oils, their fatty acids and their derivatives, efforts concentrated on preparation of novel monomers for the synthesis of materials with interesting properties, based primarily on renewable resources.

In this sense, investigations demonstrate the: i) ability of thiol-ene reactions to convert molecules derived from vegetable oils into others susceptible to polymerisation using conventional methods; and ii) possibility of using the thiol-ene reaction as a polymerisation procedure. In the latter case, the strategy is based on the synthesis of AB monomers (with thiol and ene functions as end groups) or on the coupling, *via* thiol-ene reactions, of an AA (with *thiol–thiol* end groups) and a BB (with *ene–ene* end groups) structures. The most recent and relevant studies on this subject are described in the following sections.

6.2.1 Thiol-ene Reactions for Monomer Synthesis

The first publication on the thiol-ene reaction of a vegetable-oil derivative dates to 1957, when Koenig and Swern published the free-radical addition of mercaptoacetic acid to oleic acid, methyl oleate, methyl ricinoleate and 10-undecenoic acid. That is, the synthesis of derivatives of α,ω-dicarboxylic oleic acid through thiol-ene addition [6].

The studies described here are intended to illustrate the state-of-the-art of this strategy. The works that followed this pioneering study, despite being of great importance and the foundation for other studies, are purposely omitted in this chapter.

With the main focus on modifying vegetable oils *via* the thiol-ene reaction and taking advantage of its click connotation, Bantchev and co-workers [7] undertook a systematic study of the ultraviolet-initiated thiol-ene reaction of butanethiol with canola and corn oils to produce sulfide-modified vegetable oils. Authors scanned the effect of reaction conditions such as the temperature and ratio of butanethiol:vegetable oil. This approach had been preceded by an interesting series of studies on addition of various trimercaptothiols to methyl oleate, methyl linoleate, and linseed oil to synthesise thin polymeric films on solid surfaces [8–11].

Because of the possibility of adding virtually any functional group to olefins, the thiol-ene reaction is often used as an initial step for the synthesis of new materials. This was the case in one of the first works reported by the research team of Meier, which focused on preparation of a set of 10-undecenoic acid/alcohol-derived monomers *via* the thiol-ene reaction (**Scheme 6.4**) [12] in the absence of a solvent and initiator. The thiol-ene reaction was effective and monomers were obtained at good-to-excellent yields. The ensuing α,ω-difunctional monomers, containing methyl ester and/or hydroxyl groups, were then polymerised using 1,5,7-triazabicyclo[4.4.0]dec-5-ene (TBD) as a catalyst to obtain linear and hyperbranched polyesters with number average molecular weights (Mn) of 4–10 kDa. An AB_2 monomer, used for the synthesis of hyperbranched polyesters, was prepared by reacting methyl 10-undecenoate with thioglycerol. The synthesised polyesters were stable up to 300 °C, with melting temperature (T_m) values of 50–70 °C [13].

Bao, He and Li [14] also prepared the AB_2 monomer illustrated in **Scheme 6.4**, and undertook a systematic study on the synthesis of hyperbranched polyesters using $Ti(OBu)_4$, Sb_2O_3 and $Zn(OAc)_2$ as transesterification catalysts. The final materials had Mn values of 11–60 kDa.

The research team of Cramail adopted a similar strategy to prepare aliphatic diols containing monoester, diester, ester-amide, monoamide and diamide linkages from methyl 10-undecenoate derivatives [15]. Resulting diols were condensed with a methyl diester, obtained by the self-metathesis of methyl 10-undecenoate, in the presence of TBD as a transesterification catalyst. This strategy lead to various fully aliphatic polyesters and polyester-amides with Mn values of 6–19 kDa, a glass transition temperature (T_g) below room temperature (RT) and T_m of 22–127 °C.

Another recent study on polyester synthesis through thiol-ene reactions for preparation of monomers from fatty-acid derivatives was described by Pang and co-workers [16]. Authors adopted the same approach for preparing aliphatic diols and diester from 10-undecen-1-ol, methyl 10-undecenoate and thiols. In parallel, they prepared aromatic diesters from methyl vanillate and a series of thermoplastic polyesters were synthesised by polycondensation of the diols and diesters using conventional transesterification methods. These materials were obtained with Mn values of 12–27 kDa and T_g values of –13 to 13 °C.

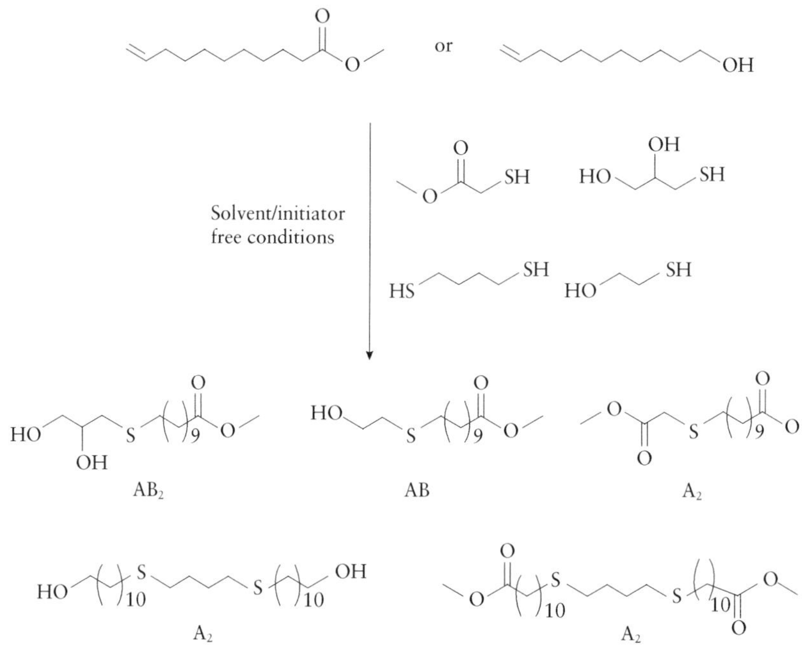

Scheme 6.4 Thiol-ene reactions leading to fatty-acid based monomers for polyester synthesis [12, 13]

Gandini and co-workers described a unique 'double click' strategy related to the preparation of monomers based on vegetable-oil derivatives bearing furan heterocycles appended through thiol-ene click chemistry, and their subsequent polymerisation *via* the Diels–Alder (DA) polycondensation between furan and maleimide complementary moieties (i.e., a second type of click chemistry). Details about the DA reaction, its mechanism, applications and the reason why it is classified as a click reaction can be found in **Chapter 7**.

In a first article [17], the authors described the synthesis of AA (with furan–furan end groups) and AB (with furan–maleimide end groups) monomers using the thiol-ene reaction between 10-undecenoic acid and 2-furanmethanethiol. Monomers bearing two terminal furan rings were reacted with bismaleimides, and monomers bearing furan and maleimide end groups were self-polymerised as described in **Chapter 7**.

A second study [18] consisted of the use of 10-undecenoic acid to synthesise potential precursors for non-linear DA reaction polycondensation. Three approaches were described in this work: i) a bisfuran monomer in combination with a trismaleimide (A_2 + B_3 system); ii) a trisfuran monomer in conjunction with a bismaleimide

(A_3 + B_2 system), leading to branched or crosslinked materials; iii) monomers incorporating furan and maleimide end groups (A_2B or AB_2 structures), which led to hyperbranched structures. In all cases, the thiol-ene reaction between 10-undecenoic acid and 2-furanmethanethiol was the main strategy to prepare the desired monomers. Corresponding DA polymerisations are discussed in **Chapter 7. Scheme 6.5** illustrates the synthetic procedures adopted for all these monomers.

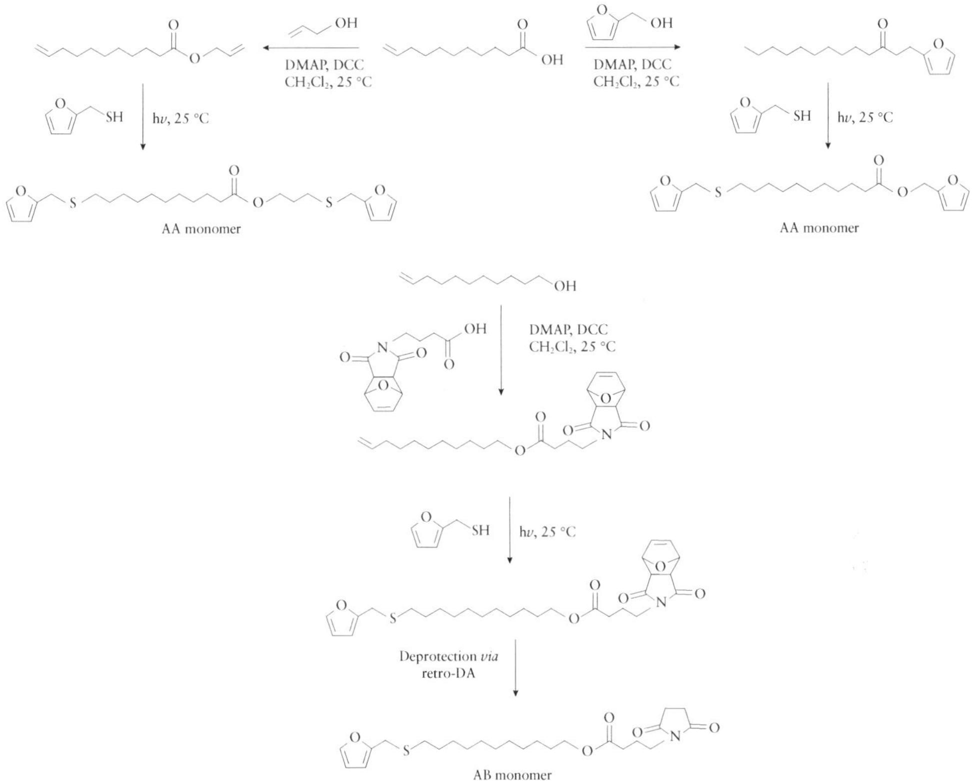

Scheme 6.5 Monomers synthesised *via* thiol-ene reaction to be used further in reversible Diels–Alder polymerisations. DCC: *N*,*N*'-Dicyclohexylcarbodiimide and DMAP: 4-dimethylaminopyridine [17, 18]

Despite the possibility of adding any organic function to C=C double bonds *via* the thiol-ene reaction, most studies on modification of fatty acids using this approach have a final goal of preparation of polyurethanes (PU), one of the most important materials in the plastic industry. Therefore, studying PU synthesised from plant oils is very important, even if it is limited only to the synthesis of polyols from fatty acids and/or triglycerides to be polymerised further with fossil-based diisocyanates [19].

Cádiz and co-workers [20] used this strategy with oleic acid and undecylenic acid. Diols obtained from their thiol-ene functionalisation were polymerised with 4,4'-methylene diphenyl diisocyanate (MDI) in *N,N*-dimethylformamide solution using tin (II) 2-ethylhexanoate as a catalyst. Polymers exhibited molar masses of 50–70 kDa and polydispersities ranging of 1.6–1.9. Amorphous (T_g = 8–20 °C) to semi-crystalline PU (T_m = 124 °C) were obtained depending on the fatty-acid derivative used [20, 21]. A similar approach was used by Boutevin and co-workers [22]. They prepared diester diols by transesterification of methyl oleate from rapeseed oil with ethylene glycol. This was followed by grafting of 2-mercaptoethanol onto the double bonds using the thiol-ene reaction with the goal of synthesising renewable PU *via* polycondensation with MDI. The obtained PU showed a T_g of –3 °C.

Also, following this same approach, the research team of Cramail reported a series of novel fatty acid-based monoester- and diester-containing diols synthesised from sunflower and ricin oils using a series of reactions (including the thiol-ene addition). In sequence, the polycondensation reactions of the diols with isophorone diisocyanate, in the presence of dibutyltin dilaurate as the catalyst, were carried out and led to materials with Mn values ≤35 kDa [23].

Desroches and co-workers [24] used a similar strategy to synthesise ester- and amide-containing pseudo-telechelic diols of fatty-acid derivatives *via* two-step reactions (i.e., transesterifications with diols, followed by thiol-ene radical coupling, and amidification reactions followed by thiol-ene radical coupling, respectively) to vary the spacer length between ester or amide functions. PU were then obtained by reaction with MDI.

The research team of Cramail demonstrated, in two very interesting studies, the possibility of synthesising AB-type monomers *via* a thiol-ene click reaction for preparation of renewable PU [25, 26]. The first described the synthesis of AB monomers using oleic acid and undecylenic acid. Their subsequent self-polycondensation was tested in the presence and absence of a catalyst. Formation of linear PU was confirmed by analyses of matrix-assisted laser desorption/ionisation-time-of-flight mass spectrometry, and their Mn values were 1–12 kDa. All PU exhibited fair thermal stability with no significant weight loss below 200 °C. The second study evaluated application of this approach to methyl oleate and methyl 10-undecenoate. In this case, the PU displayed a good thermal stability, with $T_{5\%}$ in the range of 253–295 °C, with no or negligible weight residues at 700 °C.

Renewable diisocyanates can be synthesised for the preparation of PU using the thiol-ene click reaction. This approach was first described by the research team of Cramail [27]. They reacted the prepared diisocyanates with commercially available and fatty acid-based diols to obtain partially and fully bio-based PU. These materials were obtained with values of weight average molecular weight of 6–38 kDa and displayed fair thermal stability with no significant weight loss below 235 °C. **Scheme 6.6** shows the approach adopted in this study for the synthesis of the novel diisocyanate.

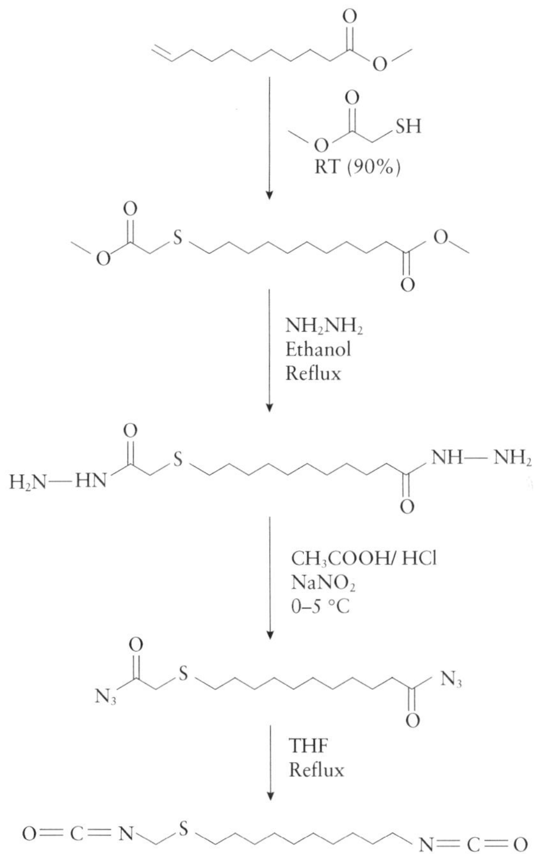

Scheme 6.6 Synthesis of a renewable diisocyanate *via* the thiol-ene click reaction. THF: Tetrahydrofuran [27]

More recently, Fu and co-workers [28] applied the same approach reported by the research team of Cramail for the synthesis of a diisocyanate, which was used as the starting monomer for the preparation of a fully bio-based waterborne PU dispersion. The material presented good thermal and hydrophobic properties.

Polyanhydrides are another important class of biodegradable polymers, offering, for instance, fast degradation properties for controlled drug-delivery applications [19]. In a publication by Cádiz and co-workers, linear and branched polyanhydrides were obtained by melt condensation of novel di- and tri-carboxylic acid monomers based on oleic acid and undecylenic acid synthesised using photo-initiated thiol-ene click chemistry. 3-Mercaptopropionic acid was added to oleic acid and 10-undecenoic acid (**Scheme 6.7**), as well as to the undecenoic acid triglyceride. This strategy resulted in di- and tri-functional derivatives of carboxylic acids, which were first acetylated and finally melt-polymerised at higher temperatures under vacuum. The polyanhydride synthesised from 10-undecenoic acid exhibited values of T_g and T_m of −38 °C and 72 °C, respectively [19, 29].

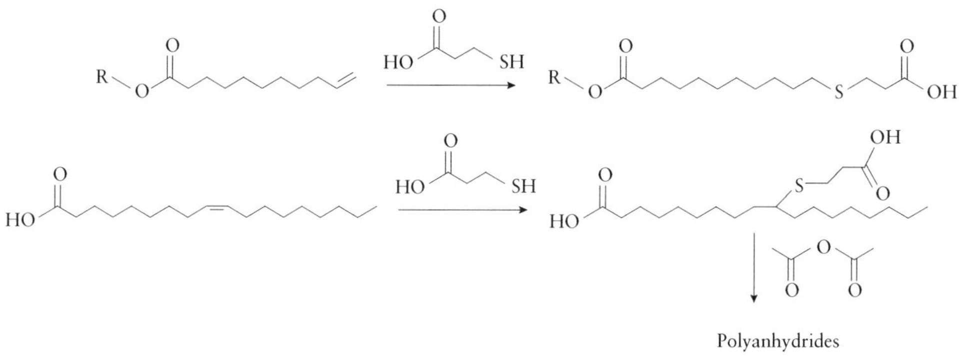

Scheme 6.7 Carboxyl monomers from sunflower oil and castor oil [19, 29]

Renewable polyamides (PA) can also be prepared taking advantage of the synthesis of thiol-ene monomers. Meier and co-workers [30] studied the thiol-ene addition of cysteamine hydrochloride to the double bonds of methyl 10-undecenoate, methyl oleate, and methyl erucate (**Scheme 6.8**), which yielded heterodifunctional monomers at good yields. Ensuing monomers were homopolymerised, copolymerised with each other, as well as copolymerised with adipic acid and 1,6-hexamethylene diamine to yield copolyamides with TBD as the catalyst. The produced PA were obtained with Mn values of 4–19 kDa.

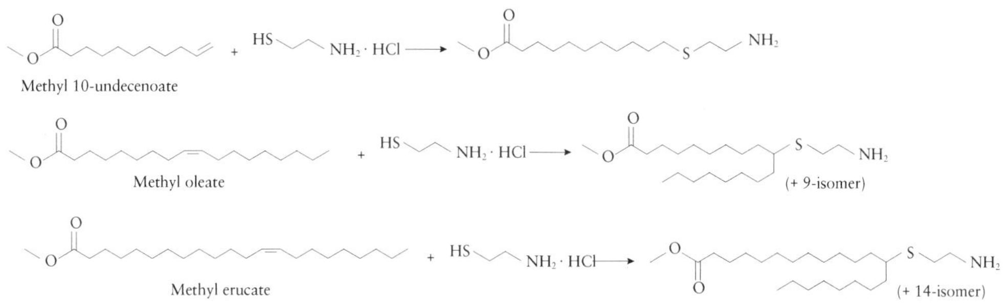

Scheme 6.8 Synthesis of new fatty acid-derived monomers for the preparation of polyamides [30]

Stemmelen and co-workers [31] used the same approach to prepare crosslinked materials from grapeseed oil. The direct thiol-ene reaction of the triglyceride with cysteamine chloride led to an aminated species with conversions ≤87% and an average number of amine units per triglyceride molecule ≈4. This molecule was reacted further with epoxidised linseed oil (**Scheme 6.9**). Analyses by differential scanning calorimetry showed a crosslinking enthalpy ≈50 $J.g^{-1}$. This value is in accordance with oil-based epoxy resin cured with diamine, which generally display crosslinking enthalpies ≈100 $J.g^{-1}$ [32].

Application of the of thiol-ene addition reaction to unsaturated plant oils and their fatty-acid derivatives represents an interesting possibility of developing technologies for the production of sustainable chemicals. In this sense, the synthesis of monomer structures with intrinsic potential to be applied to the design of new materials using this approach is very important. However, this reaction is also used as a direct polymerisation method, and its application in vegetable oils is well-established. The next section aims to illustrate the most recent advances reported on this subject.

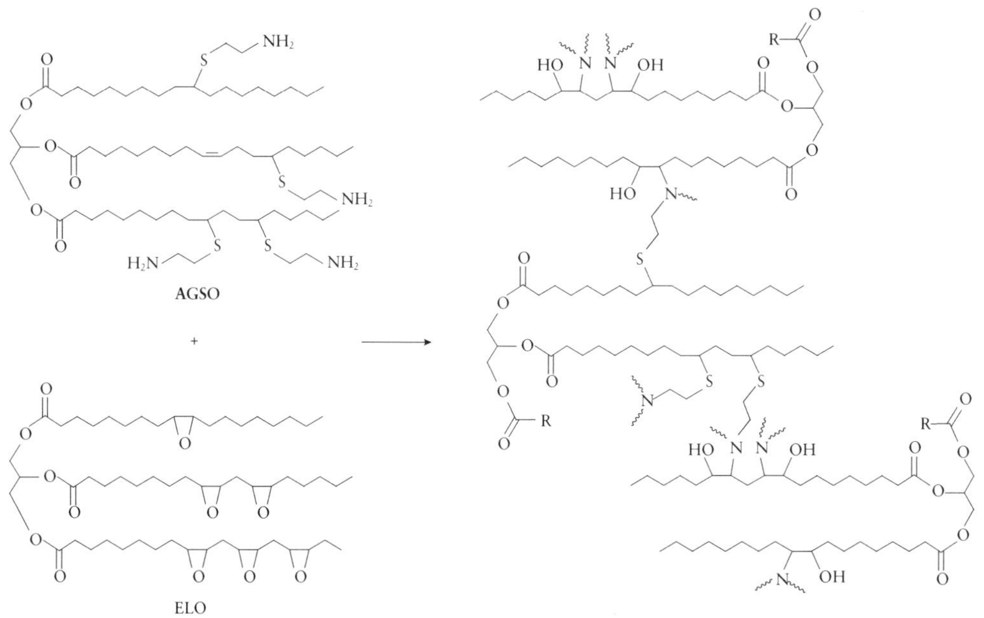

Scheme 6.9 Reaction of aminated grapeseed oil with epoxidised linseed oil. AGSO: Amino-grafted soybean oil and ELO: epoxidised linseed oil [31]

6.2.2 Thiol-ene Reactions Applied to Polymer Synthesis

Application of the thiol-ene click reaction as a direct polymerisation tool consists of the synthesis of monomers in the form of α,ω-dienes that are photo-polymerised with dithiol compounds (**Scheme 6.10a**). That is, the coupling, *via* thiol-ene reactions, of an AA (with ene–ene end groups) and a BB (with thiol–thiol end groups) structure or, alternatively, in the synthesis of AB monomers (with ene and thiol functions as end-groups, **Scheme 6.10b**). Studies reporting on the latter approach are scarce.

The mechanism of this type of polymerisation reaction follows typical step-growth polymerisation rules, and the diene:dithiol (AA:BB) ratio is therefore a very important parameter [19]. This observation implies that, in contrast to some of the grafting reactions mentioned above, the dithiol compound cannot be used in excess to compensate for the lower reaction rates with internal alkenes. Hence, these polymerisations are limited to the use of structures bearing terminal double bonds [19]. If this rule is applied to vegetable oils, 10-undecenoic acid and its derivatives are the only species that fulfill this criterion.

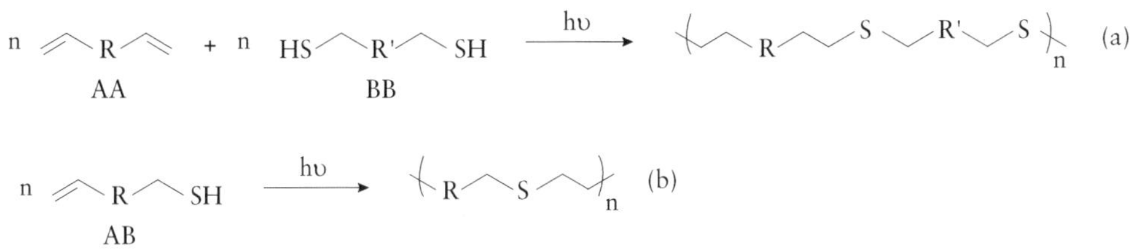

Scheme 6.10 General thiol-ene polymerisation

However, these systems offer some important advantages when compared with traditional free-radical polymerisation processes: simpler polymerisation kinetics; milder reaction conditions; insensitivity to oxygen inhibition [2, 3]. These characteristics have made the thiol-ene photo-polymerisation reaction the focus of extensive research, and its application in oleochemistry is of considerable importance in terms of polymer science and technology.

The thiol-ene addition reaction associated with oleochemistry is a very versatile tool for the polymerisation of α,ω-diene monomers bearing, for example, ester, ether, and anhydride functional groups in the main chain. Due to the availability of specific monomers, the AA/BB approach is the choice for conducting polymer syntheses most of the time.

In the first example, the research team of Meier described a route to the development of α,ω-dienes with ester and anhydride linkages (**Scheme 6.11**). These were polymerised further *via* acyclic diene metathesis (ADMET) polymerisation and thiol-ene click chemistry to compare the two processes for polyester/polyanhydride syntheses [33].

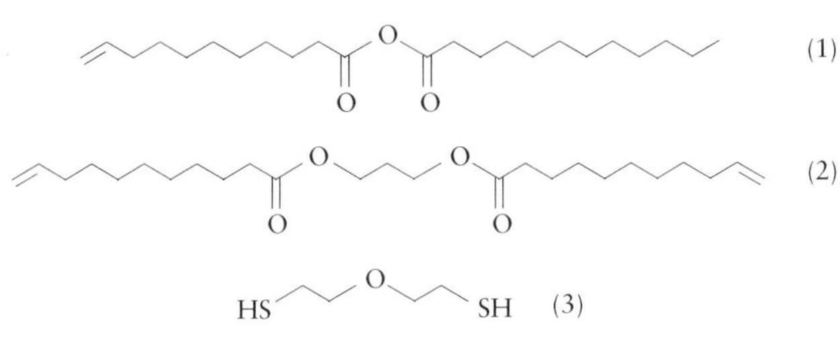

Scheme 6.11 Ene and thiol monomers for the synthesis of polyesters and polyanhydrides [33]

Authors stated that 2.5 mol% of azobisisobutyronitrile (AIBN) was used as an initiator to obtain complete conversion of the monomer after 2 h, at 80 °C. Low temperatures are favourable for thiol-ene additions. However, for polymerisations, the reaction temperature should be increased to avoid crystallisation of the polymer during polymerisation [33]. Gel permeation chromatography (GPC) analyses showed that fatty acid-based polyesters with Mn = 12 kDa (reactions **2** and **3**, **Scheme 6.11**) could be synthesised *via* thiol-ene polymerisation. However, this approach was less suitable for the polymerisation of reactions **1** and **3**, **Scheme 6.11**, which led to a Mn value of only 5 kDa. The high reactivity of the anhydride functionalities towards nucleophiles (in this case the thiol groups) caused the scission of the monomer or polymer backbone *via* thioester formation, leading to a decrease in molecular weight [33].

The same research team used α,ω-dienes derived from oleic acid and erucic acid to be copolymerised further with a derivative of ferulic acid obtained from rapeseed cake (**Scheme 6.12**). Methyl erucate and methyl oleate were epoxidised and subsequently reacted with allyl alcohol in acidic conditions to yield the corresponding α,ω-dienes [21, 34]. These monomers were polymerised by ADMET or thiol-ene reactions for the sake of comparison. For thiol-ene reactions, 1,4-butanedithiol was used to give polyhydroxyesters with Mn values of 4–14 kDa and T_g values ranging from –63 to –33 °C.

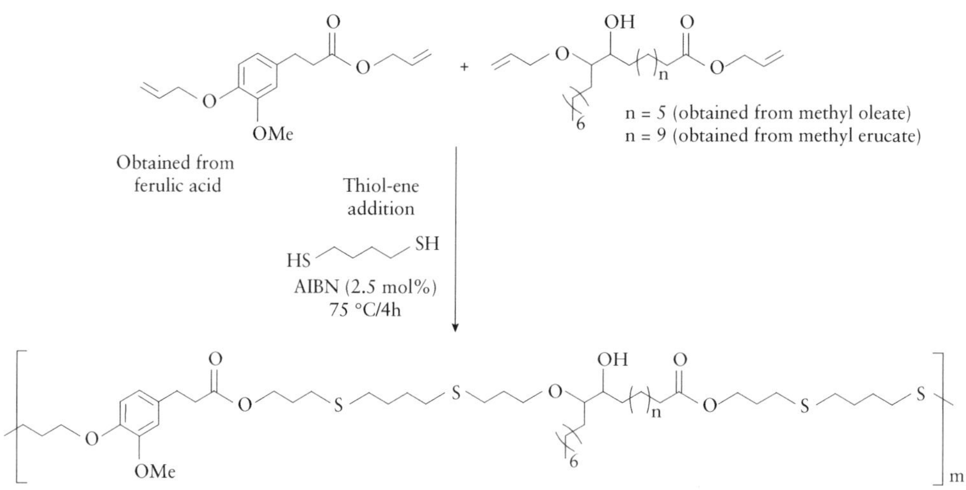

Scheme 6.12 Copolymerisation of different ratios of derivatives of fatty acids and a derivative of ferulic acid *via* the thiol-ene addition with 1,4-butanedithiol [34]

Firdaus and Meier [35] investigated the synthesis of monomers derived from vanillin and fatty acids to be used in a comparative study of ADMET, thiol-ene addition, and polycondensation-type polymerisation reactions. The highest value of Mn was obtained with the ADMET polymerisation (50 kDa). However, the thiol-ene approach also led to materials with good values: Mn = 10–16 kDa, T_g ranging from –37 to –22 °C, and T_m = 36–68 °C.

PA can also be obtained *via* the thiol-ene click reaction from α,ω-diene monomers bearing amide groups in the main backbone or as side moieties *via* the Ugi four-component reaction [36, 37]. In one such study [38], the authors used 10-undecenoic acid as carboxylic acid, 10-undecen-1-al as an aldehyde, and different primary amines and isonitriles to synthesise α,ω-diene monomers, and then applied them to polymerisation reactions. The main focus of this specific work was the ADMET mechanism, though thiol-ene photopolymerisation was also implemented to prepare linear aliphatic PA with Mn values of 3–9 kDa and T_g values of 1–10 °C.

An interesting study on the synthesis of polyethylene (PE) mimics derived from fatty-acid derivatives and thiol-ene polymerisation [39] involved preparation of two different diene monomers from a commercially available derivative of undecenoic acid, 11-bromo-1-undecene, and 10-undecenol for the synthesis of a series of non-biodegradable polymers (**Scheme 6.13**). The authors considered these non-biodegradable polymers to be renewable plant oil-derived substitutes for PE.

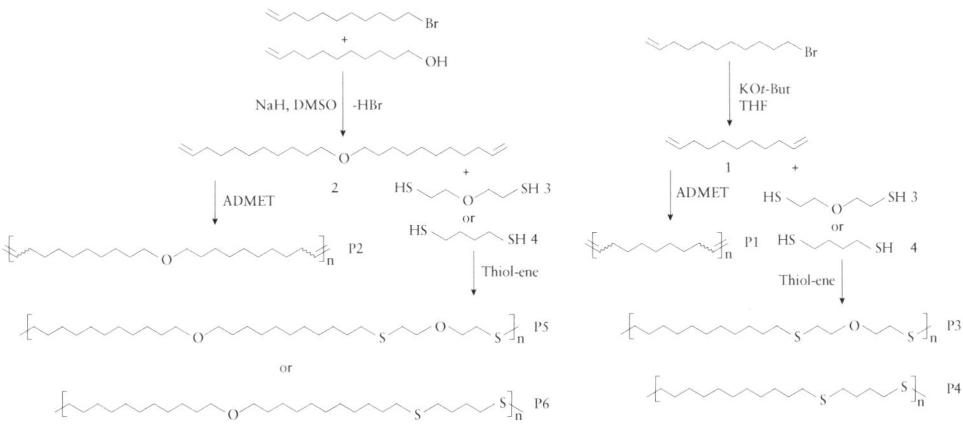

Scheme 6.13 Monomers and polyethylene-like polymers derived from castor oil. DMSO: Dimethyl sulfoxide [39]

GPC analyses of obtained materials showed higher molecular weights for thiol-ene polymers than for ADMET counterparts. The highest Mn obtained with the ADMET synthesis was 11 kDa, whereas the highest value for the thiol-ene polymer was 24 kDa, suggesting a higher efficiency of thiol-ene polymerisations for these particular monomers.

Polythioethers were also prepared with a terpene-based thiol and a long-chain diene ether [40]. 10-undecenoate was used as a platform for preparation of a renewable diene (**Scheme 6.14a**). A first dithiol (**Scheme 6.14b**) was synthesised from this diene, and a second from limonene (**Scheme 6.14c**).

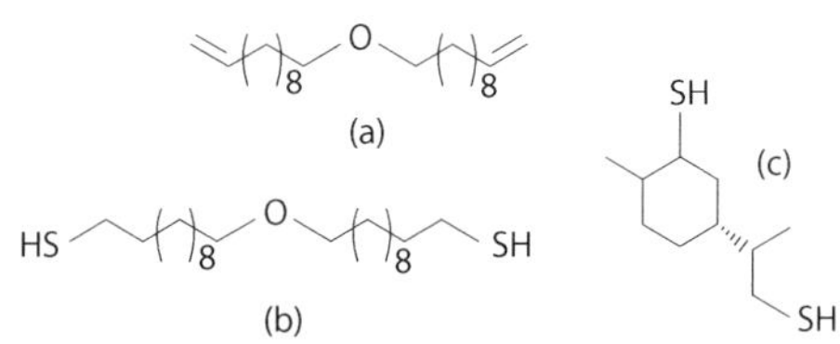

Scheme 6.14 Monomers synthesised for the preparation of renewable polythioethers [40]

Thiol-ene polymerisation reactions with monomers a + b, and a + c were carried out. The former combination led to a polymer with Mn = 32 kDa and a T_m near RT. The material formed with the latter combination was insoluble in common solvents for GPC and nuclear magnetic resonance (NMR), so determination of molecular weight was not possible.

The research team of Cramail also carried out a polythioether synthesis but used a different approach. They undertook the synthesis of AB-type monomers and then polymerised them *via* the thiol-ene click reaction [41, 42]. The work consisted of the synthesis of 10-undecene-1-thiol from 10-undecenoic acid. Photochemical or thermal initiations were tested for self-polymerisation of this AB monomer, and also by varying the reaction time. Polymers with Mn = 15–40 kDa were obtained and oxidised further, leading to materials with an increased T_m [21].

6.2.3 Thiol-ene Reactions for Chemical Modifications after Polymerisation

Post-polymerisation modifications of polymer structures based on the thiol-ene reaction are important tools in macromolecular science because several thiol compounds bearing diverse functional groups can be introduced easily onto a polymer chain that bears non-activated mono-substituted double bonds as functional groups [19].

Kolb and Meier [43] prepared a malonate derivative of methyl 10-undecenoate, which was polymerised further with 1,6-hexanediol using titanium (IV) isopropoxide as a catalyst. This polymalonate, bearing a C9 aliphatic side chain with terminal double bonds, was then subjected to grafting by ruthenium-catalysed cross-metathesis reactions with acrylates or thiol-ene addition reactions. This functionalisation enabled a subsequent Passerini multi-component reaction [44] using the pendant carboxylic-acid moiety of the modified polymers that resulted from the thiol-ene addition of 3-mercaptopropionic acid into the initial double bonds of the polymer.

Heise and co-workers [45] studied the thiol-ene functionalisation of a polyester synthesised by the enzymatic ring-opening polymerisation (ROP) of globalide (an unsaturated macrolactone prepared from hydroxyl fatty acids) (**Scheme 6.15**). Authors claimed that this was the first example of the thiol-ene functionalisation of a fatty acid-derived polymer.

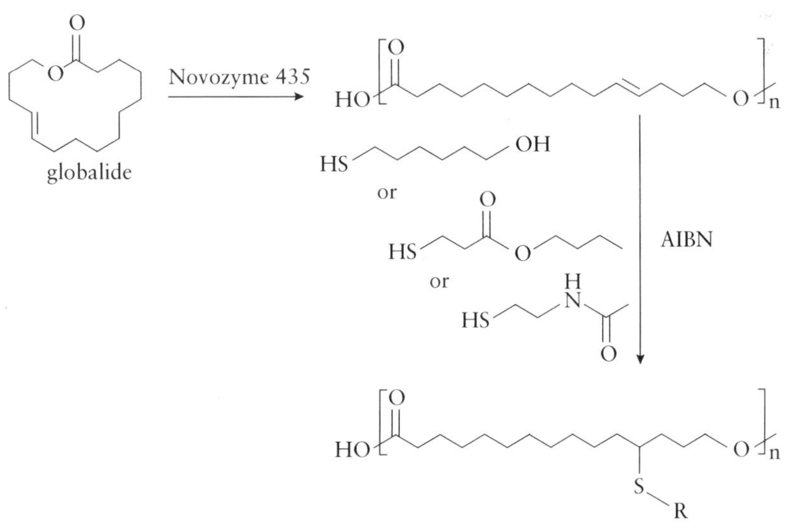

Scheme 6.15 Enzymatic polymerisation of globalide followed by thiol-ene post-polymerisation modification [45]

Quantification of the thiol-ene coupling was carried out by ^{1}H-NMR and revealed, in all cases, reduction of the double-bond peak at ≈5.4 ppm in the spectra of the final product. This finding suggested that the efficiency of thiol-ene coupling was >75% for the reaction undertaken with the thiol with an internal ester moiety, and >95% for reactions carried out with the aliphatic thiol incorporating an internal amide moiety. Values of molecular weight after functionalisation were 16–26 kDa.

Cádiz and co-workers [46] prepared telechelic polyester diols through two subsequent thiol-ene reactions. That is, a step growth polymerisation using allyl-10-undecenoate and 3,6-dioxa-1,8-octanedithiol, followed by end-group post-polymerisation modification with 2-mercaptoethanol. Telechelics based on divinyl monomers are interesting materials with potential applications in the 'tailoring' of polymer structures and properties.

Poly(2-oxazolines) are synthesised by cationic ROP of 2-R-2-oxazoline (ROx) monomers and are considered pseudo-peptides or bio-inspired polymers [21, 47]. Schubert and co-workers described the bulk polymerisation (and bulk copolymerisation with 2-ethyl-2-oxazoline) of 2-(dec-9-enyl)-2-oxazoline (DecEnOx), a fatty acid-derived monomer. In sequence, the pendant alkene groups of DecEnOx-based polymers were modified efficiently by thiol-ene reactions with dodecanethiol and 2,3,4,6-tetra-O-acetyl-1-thio-b-D-glycopyranose (**Scheme 6.16**).

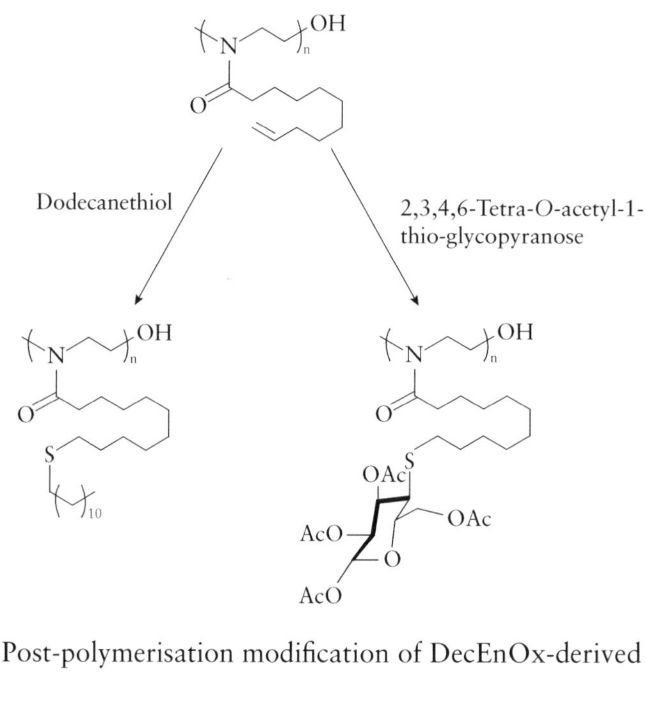

Scheme 6.16 Post-polymerisation modification of DecEnOx-derived polymers [47]

Cádiz and co-workers [48] synthesised two ROx monomers, 2-nonyl-2-oxazoline and 2-(9-decenyl)-2-oxazoline, from undecenoic acid and decanoic acid, respectively, and then copolymerised them with different feed ratios by cationic ROP. Thiol-ene addition of 2-mercaptoethanol was used to produce a set of polyoxazoline-polyols with quantitative transformation of double bonds. Polyols obtained in this way were reacted with methylene-bis(phenylisocyanate) to yield a series of amorphous and semi-crystalline PU networks.

6.3 Thiol-yne Reaction as a Tool for the Synthesis of Monomers and Polymers Derived from Vegetable Oils

Similar to the thiol-ene reaction, the thiol-yne reaction is also classified as click. Therefore, it has attracted the attention of many researchers interested in applying more environmentally friendly approaches to their work.

The mechanism of the *thiol-yne* reaction resembles that of its *thiol-ene* sister. However, a second possible cycle involving further addition of a thiol into the remaining double-bond reaction is also possible in the thiol-yne version (**Scheme 6.17**) [49]. As a matter of fact, this second cycle is, by definition, a thiol-ene radical coupling.

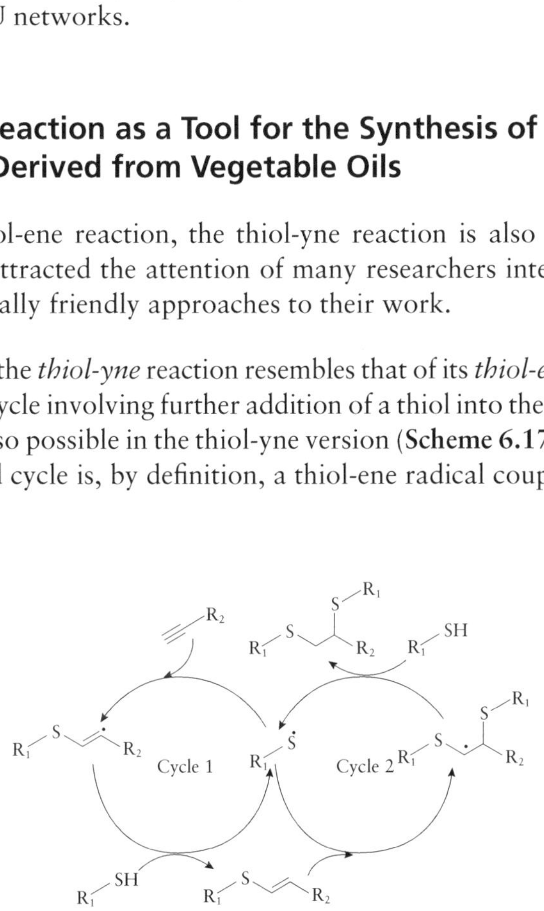

Scheme 6.17 Radical-mediated mechanism of the thiol-yne click reaction [49]

In the first cycle (**Scheme 6.17**), the thiyl radicals are generated from thiol groups by photo-initiation or thermo-initiation. The thiyl radical then adds across the ethynyl group to form a vinyl sulfide radical that abstracts a hydrogen atom from a thiol group to produce the vinyl sulfide and another thiyl radical simultaneously. In the second cycle (**Scheme 6.17**), which can occur or not depending on system conditions, a thiyl radical adds across the double bond of the vinyl sulfide to generate a dithioether radical that abstracts a hydrogen atom from a thiol group and forms the disubstituted product and a new thiyl radical [49]. As with the thiol-ene reaction, thiol-yne addition generally proceeds rapidly under various experimental conditions to selectively yield mono- or bis-addition products [50].

There are many reports on azide-alkyne click polymerisation (a subject not discussed in the present chapter) but research on thiol-yne click polymerisation is in its early stages. New reaction types, novel catalyst systems other than those existing (i.e., photon, heat, organic base, transition–metal complexes) and new functionalities of the ensuing polymers are waiting to be developed [51].

Only a few studies on oleochemistry have been reported until now. Similar to the effects seen with the thiol-ene reaction, one can take advantage of this interesting tool for the synthesis of novel monomers to be polymerised further by classical methods. Also, the possibility of two thiol groups reacting with one ethynyl group *via* two-step addition makes thiol-yne click polymerisation an interesting method for the preparation of hyperbranched polymers.

Despite being 'sister reactions', works reporting on use of thiol-yne reactions for the synthesis of monomers and polymers from fatty-acid derivatives are much scarcer than those using the thiol-ene approach.

Cádiz and co-workers [50] demonstrated use of the thiol-yne reaction for the synthesis of polyols for application in partially renewable PU (**Scheme 6.18**). The starting materials were 10-undecynoic acid and 9-octadecynoic acid, which were prepared by dehydrobromination of the corresponding derivatives of 10-undecenoic acid and 9-octadecenoic acid.

Diols depicted in **Scheme 6.18** were combined with MDI in dimethylformamide solution at 50 °C for 24 h using tin(II) 2-ethylhexanoate as the catalyst to prepare these bio-based PU. Triols were used to prepare PU thermosets by reaction with MDI in THF. Linear PU materials showed T_g values just below RT and no crystallinity. Crosslinked PU displayed higher T_g values (42–59 °C). In addition, authors evaluated the biocompatibility of the synthesised PU for tissue-engineering applications in orthopedic/trauma surgery.

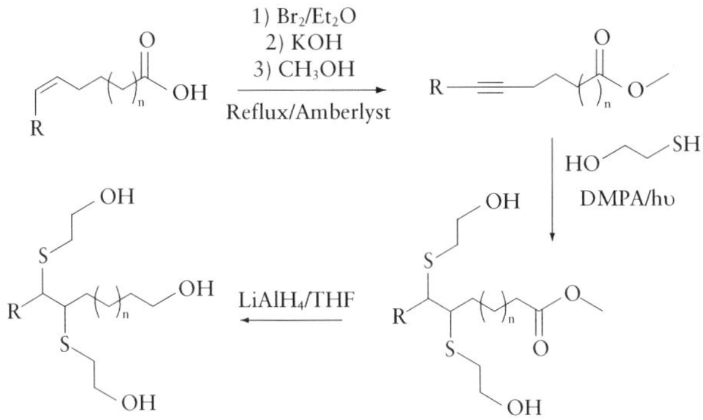

Scheme 6.18 Synthesis of fatty acid-derived polyols starting from 10-undecenoic acid and oleic acid. DMPA: 2,2-Dimethoxy-2-phenylacetophenone [50]

A very similar strategy was adopted recently by Pham and co-workers [52] when propargylic fatty ester (PFE) and propargylic fatty diester (PFD) derived from propargyl alcohol and a derivative of oleic acid were converted into lipidic polyols (OH1) and (OH2), respectively, *via* thiol-ene and thiol-yne reactions with mercaptoethanol (**Scheme 6.19**).

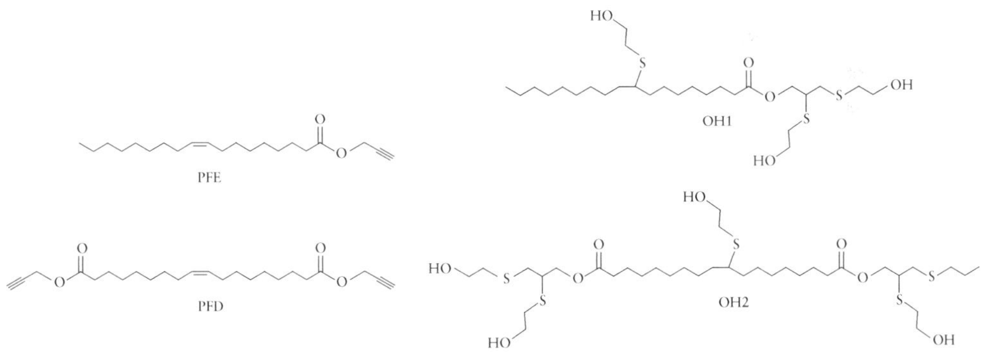

Scheme 6.19 Application of thiol-ene and thiol-yne reactions of propargylic fatty ester and propargylic fatty diester with mercaptoethanol [52]

As in the study conducted by Cádiz and co-workers [50], authors used the ensuing polyols for PU production with MDI. The T_g of these materials was 36–72 °C depending on the number of alcohol groups of each fatty chain.

Türünç and Meier [53] described direct polymerisation of monoalkynes using thiol-yne addition. First, an optimisation step was conducted with 1-octyne and octanethiol as model compounds in the absence of an initiator or using thermal (AIBN) or photochemical initiation of the DMPA radical. Optimal conditions were determined and authors could evaluate the ability to prepare a range of functional polymers using radical thiol-yne addition with dithiols. Polymerisation of several functional monoalkynes, such as 10-undecynoic acid and propargylic acid, was investigated: highly functional comb-like linear polymers were accessible *via* this approach [54].

6.4 Final Considerations

The search for novel polymeric structures is the focus of a tremendous number of studies all over the world. However, just searching for new polymers is deemed unsatisfactory because environmental considerations that meet the need of society must also prevail.

In this context, vegetable oils and their fatty acids are key structures because they offer a wide range of possibilities for polymer syntheses based on renewable resources. Thiol-ene and thiol-yne click reactions applied to vegetable oils and their fatty acids constitute a promising strategy to meet those goals. This combination is relatively new and, despite the important contributions achieved, many more are needed.

References

1. T. Posner, *Berichte der Deutschen Chemischen Gesellschaft*, 1905, **38**, 646.

2. H.C. Kolb, M.G. Finn and K.B. Sharpless, *Angewandte Chemie*, 2001, **113**, 2056.

3. H.C. Kolb, M.G. Finn and K.B. Sharpless, *Angewandte Chemie International Edition*, 2001, **40**, 2004.

4. C.E. Hoyle and C.N. Bowman, *Angewandte Chemie International Edition*, 2010, **49**, 1540.

5. A.B. Lowe, *Polymer Chemistry*, 2010, **1**, 17.

6. N.H. Koenig and D. Swern, *Journal of the American Chemical Society*, 1957, **79**, 362.

7. G.B. Bantchev, J.A. Kenar, G. Biresaw and M.G. Han, *Journal of Agricultural and Food Chemistry*, 2009, **57**, 1282.

8. U. Bexell, R. Berger, M. Olsson, T.M. Grehk, P-E. Sundell and M. Johansson, *Thin Solid Films*, 2006, **2**, 838.

9. J. Samuelsson, M. Jonsson, T. Brinck and M.J. Johansson, *Journal of Polymer Science, Part A: Polymer Chemistry Edition*, 2004, **42**, 6346.

10. U. Bexell, M. Olsson, P-E. Sundell, M. Johansson, P. Carlsson and M. Hellsing, *Applied Surface Science*, 2004, **231–232**, 362.

11. U. Bexell, M. Olsson, M. Johansson, J. Samuelsson and P-E. Sundell, *Surface and Coatings Technology*, 2003, **166**, 141.

12. O. Türünç and M.A.R. Meier, *Macromolecular Rapid Communications*, 2010, **31**, 1822.

13. L. Montero de Espinosa and M.A.R. Meier, *European Polymer Journal*, 2011, **47**, 837.

14. Y. Bao, J. He and Y. Li, *Polymer International*, 2013, **62**, 1457.

15. T. Lebarbé, L. Maisonneuve, T.H.N. Nguyen, B. Gadenne, C. Alfos and H. Cramail, *Polymer Chemistry*, 2012, **3**, 2842.

16. C. Pang, J. Zhang, G. Wu, Y. Wang, H. Gaob and J. Ma, *Polymer Chemistry*, 2014, **5**, 2843.

17. C. Vilela, L. Cruciani, A.J.D. Silvestre and A. Gandini, *Macromolecular Rapid Communications*, 2011, **32**, 1319.

18. C. Vilela, A.J.D. Silvestre and A. Gandini, *Journal of Polymer Science, Part A: Polymer Chemistry Edition*, 2013, **51**, 2260.

19. O. Türünç and M.A.R. Meier, *European Journal of Lipid Science and Technology*, 2013, **115**, 41.

20. R.J. González-Paz, C. Lluch, G. Lligadas, J.C. Ronda, M. Galià and V. Cádiz, *Journal of Polymer Science, Part A: Polymer Chemistry Edition*, 2011, **49**, 2407.

21. L. Maisonneuve, T. Lebarbé, E. Grauab and H. Cramail, *Polymer Chemistry*, 2013, **4**, 5472.

22. M. Desroches, S. Caillol, R. Auvergne and B. Boutevin, *European Journal of Lipid Science and Technology*, 2012, **114**, 84.

23. D.V. Palaskar, A. Boyer, E. Cloutet, J-F. Le Meins, B. Gadenne, C. Alfos, C. Farcet and H. Cramail, *Journal of Polymer Science, Part A: Polymer Chemistry Edition*, 2012, **50**, 1766.

24. M. Desroches, S. Caillol, R. Auvergne, B. Boutevin and G. David, *Polymer Chemistry*, 2012, **3**, 450.

25. A.S. More, B. Gadenne, C. Alfos and H. Cramail, *Polymer Chemistry*, 2012, **3**, 1594.

26. A.S. More, L. Maisonneuve, T. Lebarbé, B. Gadenne, C. Alfos and H. Cramail, *European Journal of Lipid Science and Technology*, 2013, **115**, 61.

27. A.S. More, T. Lebarbé, L. Maisonneuve, B. Gadenne, C. Alfos and H. Cramail, *European Polymer Journal*, 2013, 49, 823.

28. C. Fu, Z. Zheng, Z. Yang, Y. Chena and L. Shen, *Progress in Organic Coatings*, 2014, 77, 53.

29. C. Lluch, G. Lligadas, J.C. Ronda, M. Galià and V. Cádiz, *Macromolecular Rapid Communications*, 2011, **32**, 1343.

30. O. Türünç, M. Firdaus, G. Klein and M.A.R. Meier, *Green Chemistry*, 2012, **14**, 2577.

31. M. Stemmelen, F. Pessel, V. Lapinte, S. Caillol, J-P. Habas and J-J. Robin, *Journal of Polymer Science, Part A: Polymer Chemistry Edition*, 2011, **49**, 2434.

32. A. Shabeer, S. Sundararaman, K. Chandrashekhara and L.R. Dharani, *Journal of Applied Polymer Science*, 2007, **105**, 656.

33. O. Türünç and M.A.R. Meier, *Green Chemistry*, 2011, **13**, 314.

34. O. Kreye, T. Tóth and M.A.R. Meier, *European Polymer Journal*, 2011, **47**, 1804.

35. M. Firdaus and M.A.R. Meier, *European Polymer Journal*, 2013, **49**, 156.

36. I. Ugi and C. Steinbrückner, *Angewandte Chemie*, 1960, **72**, 267.

37. I. Ugi, *Angewandte Chemie International Edition in English*, 1962, **1**, 8.

38. O. Kreye, O. Türünç, A. Sehlinger, J. Rackwitz and M.A.R. Meier, *Chemistry – A European Journal*, 2012, **18**, 5767.

39. O. Türünç, L. Montero de Espinosa and M.A.R. Meier, *Macromolecular Rapid Communications*, 2011, **32**, 1357.

40. M. Firdaus, M.A.R. Meier, U. Biermann and J.O. Metzger, *European Journal of Lipid Science and Technology*, 2014, **116**, 31.

41. O. van den Berg, T. Dispinar, B. Hommez and F.E. Du Prez, *European Polymer Journal*, 2013, **49**, 804.

42. F. Deubel, V. Bretzler, R. Holzner, T. Helbich, O. Nuyken, B. Rieger and R. Jordan, *Macromolecular Rapid Communications*, 2013, **34**, 1020.

43. N. Kolb and M.A.R. Meier, *European Polymer Journal*, 2013, **49**, 843.

44. L. Banfi and R. Riva in *Organic Reactions*, John Wiley & Sons, Inc., New York, NY, USA, 2004.

45. Z. Ates, P.D. Thorntona and A. Heise, *Polymer Chemistry*, 2011, **2**, 309.

46. C. Lluch, J.C. Ronda, M. Galià, G. Lligadas and V. Cádiz, *Biomacromolecules*, 2010, **11**, 1646.

47. H. Schlaad, C. Diehl, A. Gress, M. Meyer, A.L. Demirel, Y. Nur and A. Bertin, *Macromolecular Rapid Communications*, 2010, **31**, 511.

48. K. Kempe, R. Hoogenboom and U.S. Schubert, *Macromolecular Rapid Communications*, 2011, **32**, 1484.

49. B. Yao, J. Sun, A. Qin and B.Z. Tang, *Chinese Science Bulletin*, 2013, **58**, 2711.

50. R.J. González-Paz, G. Lligadas, J.C. Ronda, M. Galià and V. Cádiz, *Polymer Chemistry*, 2012, **3**, 2471.

51. B. Yao, J. Mei, J. Li, J. Wang, H. Wu, J.Z. Sun, A. Qin and B.Z. Tang, *Macromolecules*, 2014, **47**, 1325.

52. P.D. Pham, V. Lapinte, Y. Raoul and J-J. Robin, *Journal of Polymer Science, Part A: Polymer Chemistry Edition*, 2014, **52**, 1597.

53. O. Türünç and M.A.R. Meier, *Journal of Polymer Science, Part A: Polymer Chemistry Edition*, 2012, **50**, 1689.

54. G. Lligadas, J.C. Ronda, M. Galià and V. Cádiz, *Journal of Polymer Science, Part A: Polymer Chemistry Edition*, 2013, **51**, 2111.

7 Diels–Alder Reactions and Polycondensations Applied to Vegetable Oils and their Derivatives

The Diels–Alder (DA) reaction is one of the most basic and useful reactions in synthetic organic chemistry because it provides simple, efficient, and clean procedures to generate new bonds by intermolecular or intramolecular couplings. It was discovered in 1928 by Otto Diels and Kurt Alder, who were awarded the 1950 Nobel Prize in chemistry. Since then, it has received a great deal of attention, as shown by the numerous publications concerning its synthetic, mechanistic and theoretical aspects [1–5].

The DA reaction is a recognised member of the 'click chemistry' family because in most of its applications it complies with the requirements set out by Sharpless and co-workers [6], as discussed in **Chapter 6** in the context of the thiol-ene reaction.

The classical DA reaction is a [4+2] cycloaddition between a conjugated diene and a second component ('dienophile') to give a stable cyclohexene derivative ('adduct') [1–5]. This reaction displays a thermally reversible character, which allows decoupling of the adduct to occur by increasing the temperature. Hence, the equilibrium is displaced to the left with regeneration of initial reagents (**Scheme 7.1**). The reverse reaction is called the 'retro-Diels–Alder' (retro-DA) reaction [1–5].

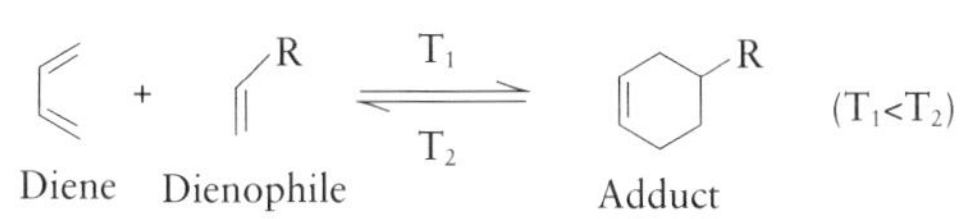

Scheme 7.1 Diels–Alder and retro-Diels–Alder reactions

With regard to polymers from renewable resources, the DA reaction has been applied mainly to the combination of furan moieties ('dienes') and maleimide complementary functions ('dienophiles') [7]. However, very recently, specific conjugated structures (dienes) present in some plant oils have also been investigated in conjunction with maleimides, as discussed below.

The furan–maleimide DA reaction is particularly useful because it provides a means to synthesise polymers through the thermo-reversible polycondensation mechanism shown in **Scheme 7.2**. That is, the forward step (chain growth) can be carried out at ≤65 °C without the need for a catalyst, whereas ≥110 °C the reverse step (chain breaking) dominates [7].

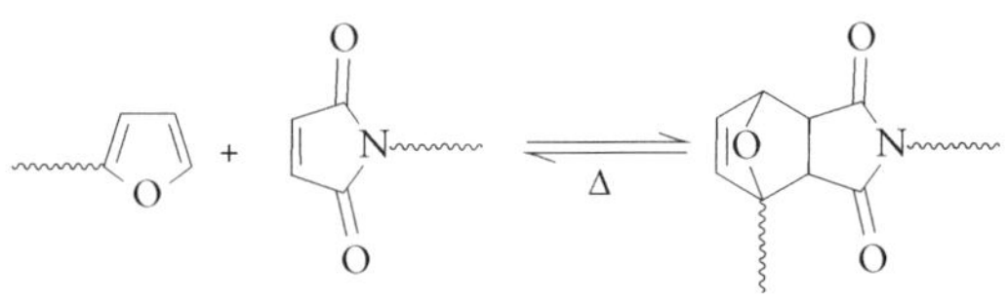

Scheme 7.2 Diels–Alder equilibrium between furan and maleimide end groups in a stepwise macromolecular synthesis

This strategy has produced a vast number of studies, particularly within the last decade [7]. There are four main approaches adopted to implement this general strategy. The first approach is linear polycondensations involving the reaction of bisfuran AA monomers with bismaleimides (BM) (BB monomers), or use of a furan–maleimide AB monomer. The second approach is non-linear architectures using complementary monomers with an average functionality >2, with classical A_3+B_2 (or *vice versa*) monomer combinations to prepare highly branched or crosslinked polymers (depending on the system stoichiometry) or with AB_2 or A_2B monomers to synthesise hyperbranched macromolecules that cannot form a network. The third approach is the crosslinking reaction of a linear polymer incorporating pendant furan heterocycles with a BM (or *vice versa*). The final approach involves other, more specific situations, such as construction of dendritic macrostructures and decoration of nanoparticles [7]. In all these approaches, the underlying feature is their reversible connotation, which enables materials to be returned to the corresponding starting reagents by simple heating (i.e., recyclability), which is particularly important for the normally intractable gelled polymers.

Implementation of these concepts to polymers based on vegetable oils [8] started with a set of studies in which 10-undecenoic structures were converted into *bis*- and *tri*-furan monomers, as well as AB, AB_2 and A_2B homologous structures bearing furan and maleimide moieties [9–11]. Syntheses of these monomers were carried out using thiol-ene reactions in conjunction with more conventional couplings, as discussed in **Chapter 6** (the mechanisms are illustrated in **Scheme 6.5**).

DA polymerisation of difuran monomers was carried out at 60 °C using different BM to prepare polymers with different glass transition temperature (T_g) values (–40 to 0 °C), together with polycondensation of the AB homologue. Structure and properties of all ensuing materials were investigated thoroughly, as well as their thermal reversibility at 110 °C [9, 10]. Their molecular weights varied from 7 to 17 kDa. **Scheme 7.3** shows one such linear DA polycondensation.

The DA synthesis (under the same conditions as for linear polymerisations) of non-linear counterparts called upon various combinations of di- and tri-functional monomers, including trismaleimides. However, self-condensation of A_2B and AB_2 homologues (see **Chapter 6**) [11] also gave various branched and crosslinked products depending on the monomers used and their initial stoichiometry. Interesting materials were obtained and characterised fully, including their thermal reversibility. One of these systems is shown in **Scheme 7.4**.

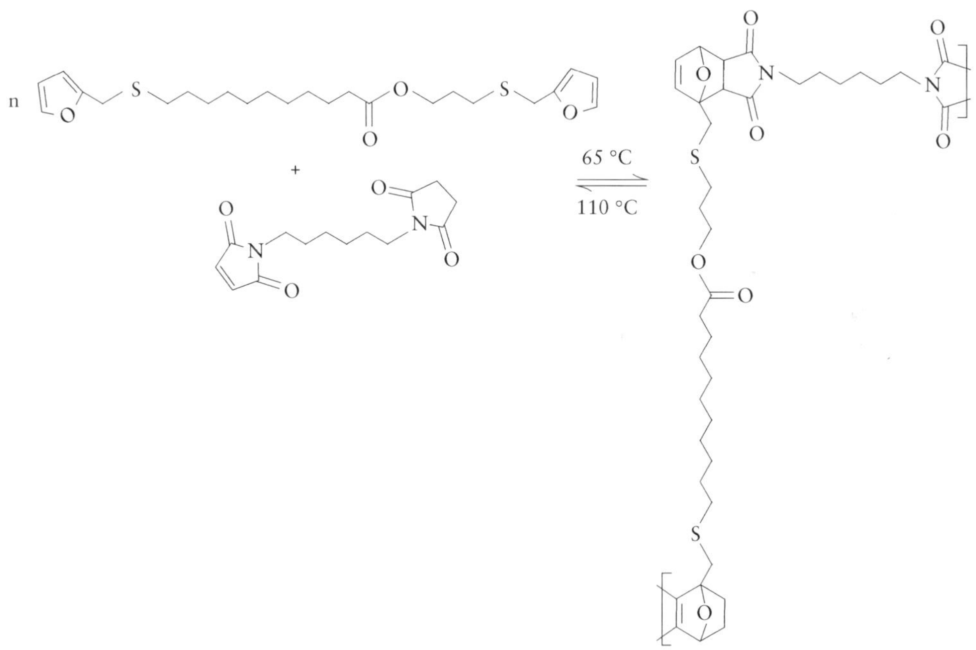

Scheme 7.3 Typical thermally reversible Diels–Alder polymerisation with a difuran monomer prepared from a derivative of vegetable oils [9, 10]

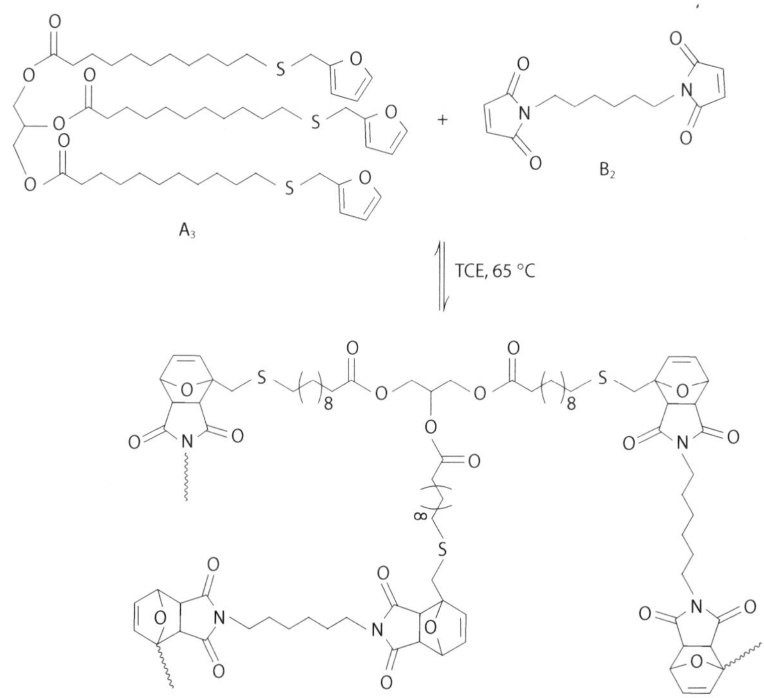

Scheme 7.4 Non-linear Diels–Alder polycondensation between an oil-based trisfuran and a BM. TCE: 1,1,2,2-Tetrachloroethane [11]

These studies opened the way to a new area of macromolecular science based on renewable resources pursued by approaches based on pristine oils and their commercial derivatives. The bulk reaction of epoxidised vegetable oils with an excess of furfurylamine was studied in 'green chemistry' conditions and found to take place through two parallel mechanisms occurring at approximately the same rate. That is, aminolysis of ester bonds and oxirane ring-opening, as shown in **Scheme 7.5** for the case of epoxidised linseed oil [12]. These reactions were followed by Fourier-Transform infrared spectroscopy and ^{1}H-nuclear magnetic resonance spectroscopy, respectively, and both reached completion.

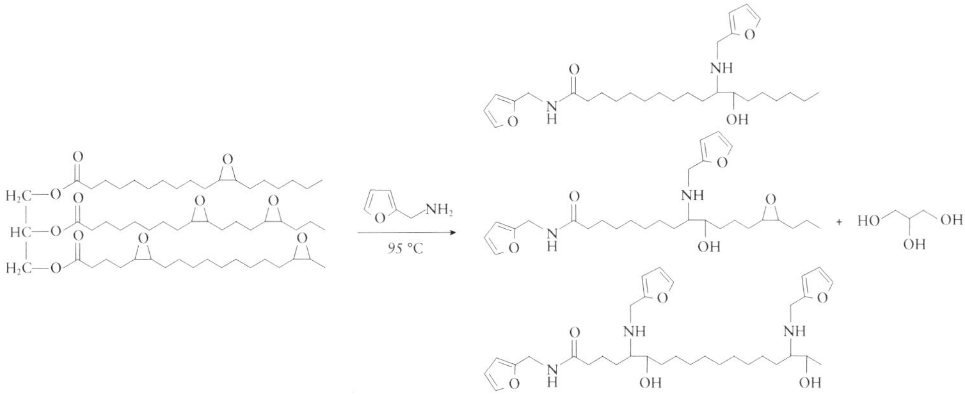

Scheme 7.5 Double insertion of furfurylamine into epoxidised linseed oil [12]

The mixture of *bis*- and *tris*-furan fatty-acid products were then DA-polymerised with BM. The corresponding polycondensations are shown in **Schemes 7.6** and **7.7**, respectively [12].

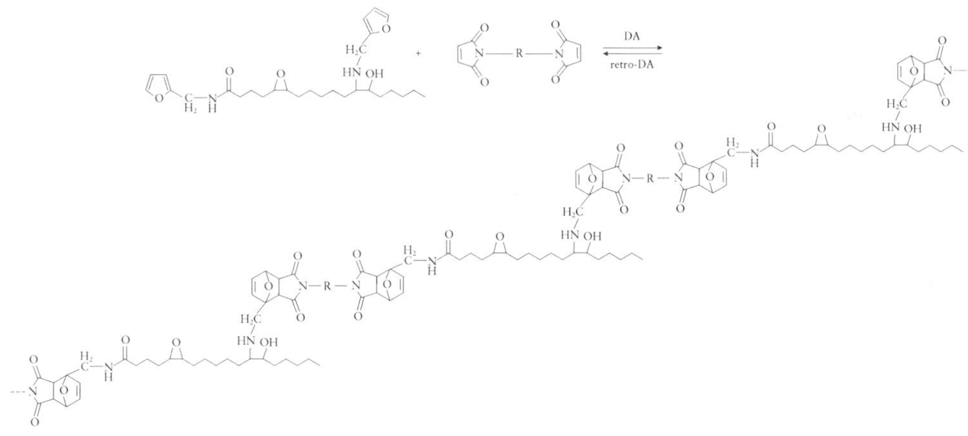

Scheme 7.6 Linear Diels–Alder polymerisation of a difuran product with an aromatic bismaleimide [12]

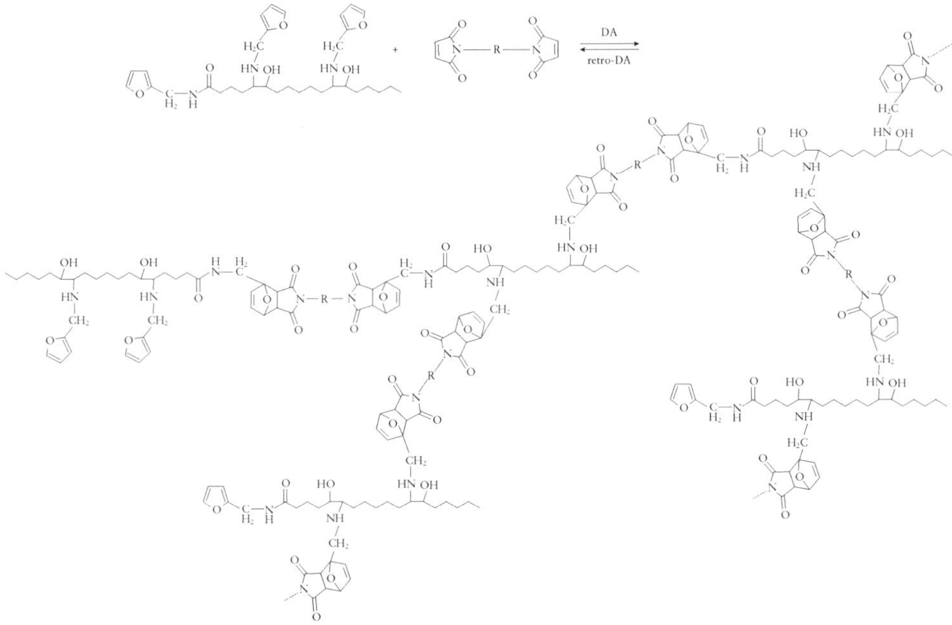

Scheme 7.7 Non-linear Diels–Alder polymerisation of the trisfuran with an aromatic bismaleimide [12]

Concomitance of the two reactions did not lead to a gelled material because the proportion of the trisfuran product was quite modest in relation to the bisfuran homologue. Molecular weights of these polymers were ≈40 kDa with polydispersity index (PDI) values ≈5, clearly indicating branched macromolecules arising from the DA reaction of the trisfuran monomer [12]. T_g values were 80–100 °C and DA thermal reversion to starting reagents took place >100 °C.

This approach was also applied to other epoxidised vegetable oils with entirely similar results, which proved its wide scope and reproducible features.

The next step in this new venture consisted in tackling pristine tung oil (TO) as the possible reagent in DA polymerisations [13]. TO is readily obtained from the seeds of the tung tree (*Vernicia fordii*) nuts. It is a relatively cheap commodity used for millennia as a source of siccative coatings. It incorporates as its main constituent (≈85%) a peculiar triglyceride structure in which each chain bears three conjugated unsaturations corresponding to α-eleostearic acid. These moieties behave like very active 'dienes' in the DA reaction, as first shown by the research team of Metzger in a study with maleic anhydride as a dienophile [14] and later in reactions of TO with methylmaleimide [8].

Two systems were devised to prepare TO-based polymers using DA coupling [13]. The first involved investigation of the reaction of the pristine oil with different BM (**Scheme 7.8a**).

These non-linear polycondensations were rather fast even when using a default of BM with respect to group stoichiometry. Ensuing gels had T_g values ranging from –10 °C to 75 °C as a function of the increasing molecular stiffness of the BM. Their thermal stability reached 350 °C. Application of this type of DA polymerisation appears to be very suitable for drying film coatings by dissolving the BM in the oil before spreading it onto the substrate. Interestingly, the DA reaction between the triene moiety of TO and a dienophile is *not* thermally reversible within reasonable temperatures because the mechanism of adduct formation implies a drastic structural change.

The second approach involved initially applying the aminolysis reaction with an excess of furfurylamine (as described above) to TO, which gave three fatty-acid amides with the triene moiety and a furan end group, and then polymerising them with the same selection of BM used in the first approach [13]. These linear polycondensations involved, therefore, two distinct DA couplings: one between a maleimide function and a furan heterocycle, and the other between a maleimide function and the triene moiety (**Scheme 7.8b**).

The T_g of these thermoplastic materials was 30–105 °C (i.e., higher than those of the crosslinked counterparts) because incorporation of amide groups in their macromolecules resulted in a decrease in chain flexibility associated with intermolecular C=O···H–N hydrogen bonds. It was, moreover, verified that their thermal reversibility was limited to furan–maleimide adducts (i.e., only half of the DA couplings), with the others being thermally stable.

A recent addition to the use of TO in DA polymerisations [15] describes a mixed system in which a TO fatty-acid derivative bearing a terminal epoxy group was variously crosslinked using maleic anhydride (which reacted by DA coupling with the triene moiety and with two epoxide end groups) or an alicyclic anhydride and a BM for separately achieving the same purpose. The T_g of these networks was 75–105 °C and their mechanical properties could be modulated by varying the crosslinking mechanism.

The most relevant contributions on the furan-maleimide DA reaction as a tool for the preparation of recyclable polymers based on triglycerides and their corresponding macromonomers, polyssacharides, natural rubber and gelatin were recently reviewed [16]. Although in its infancy, application of the DA reaction to the synthesis of novel, mostly thermo-reversible, polymers based on vegetable oils and other renewable resources is progressively gaining a respectable status, which will undoubtedly advance at a good pace in years to come.

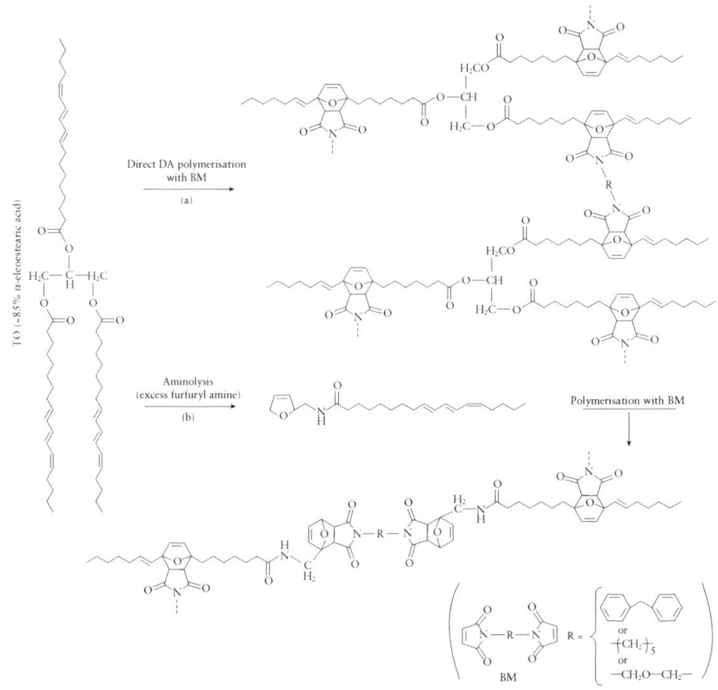

Scheme 7.8 Crosslinking of tung oil with different bismaleimide through Diels–Alder (DA) intermolecular couplings (a) and its aminolysis followed by linear DA polymerisation of the ensuing fatty acid furan amides (b) [13]

References

1. F. Fringuelli and A. Taticchi in *Dienes in the Diels–Alder Reaction*, Wiley-VCH, New York, NY, USA, 1990.

2. F. Fringuelli and A. Taticchi in *The Diels–Alder Reaction: Selected Practical Methods*, John Wiley & Sons, Chichester, UK 2002.

3. K.C. Nicolau, S.A. Snyder, T. Montagnon and A. Vassilikogiannakis, *Angewandte Chemie International Edition*, 2002, **41**, 1669.

4. E.M. Carreira and L. Kvaerno in *Classics in Stereoselective Synthesis*, Wiley-VCH, Weinheim, Germany, 2009.

5. W. Carruthers and I. Coldham in *Cycloaddition Reactions in Organic Synthesis*, Cambridge University Press, Cambridge, UK, 2004.

6. H.C. Kolb, M.G. Finn and K.B. Sharpless, *Angewandte Chemie International Edition*, 2001, **40**, 2004.

7. A. Gandini, *Progress in Polymer Science*, 2013, **38**, 1.

8. T.M. Lacerda and A. Gandini, *Journal of Renewable Materials*, 2014, **2**, 2.

9. C. Vilela, L. Cruciani, A.J.D. Silvestre and A. Gandini, *Macromolecular Rapid Communications*, 2011, **32**, 1319.

10. C. Vilela, L. Cruciani, A.J.D. Silvestre and A. Gandini, *RSC Advances*, 2012, **2**, 2966.

11. C. Vilela, A.J.D. Silvestre and A. Gandini, *Journal of Polymer Science, Part A: Polymer Chemistry Edition*, 2013, **51**, 2260.

12. A. Gandini, T.M. Lacerda and A.J.F. Carvalho, *Green Chemistry*, 2013, **15**, 1514.

13. T.M. Lacerda, A.J.F. Carvalho and A. Gandini, *RSC Advances*, 2014, **4**, 26829.

14. U. Biermann, W. Butte, T. Eren, D. Haase and J.O. Metzger, *European Journal of Organic Chemistry*, 2007, **23**, 3859.

15. K. Huang, Z. Liu, J. Zhang, S. Li, M. Li, J. Xia and Y. Zhou, *Biomacromolecules*, 2014, **15**, 837.

16. A. Gandini, A. J. F. Carvalho, E. Trovatti, R. K. Kramer and T. M. Lacerda, *European Journal of Lipid Science and Technology*, 2018, **120**, 1700091.

Index

Made in the USA
Columbia, SC
01 October 2026

c7e2e8d4-4ff5-4681-8a5d-e1fa89a2bc1cR01